Oxford Revise

Revision & Practice

AQA GCSE 9–1
CHEMISTRY
HIGHER

Knowledge Retrieval Practice

Series Editor: **Primrose Kitten**

Adam Boxer

Philippa Gardom Hulme

OXFORD
UNIVERSITY PRESS

Contents

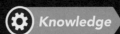  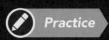

 Shade in each level of the circle as you feel more confident and ready for your exam.

For answers and more practice questions visit
www.oxfordrevise.com/scienceanswers

Even more practice and interactive
revision quizzes are available on kerboodle

How to use this book

This book uses a three-step approach to revision: **Knowledge**, **Retrieval**, and **Practice**. It is important that you do all three; they work together to make your revision effective.

1 Knowledge

Knowledge comes first. Each chapter starts with a **Knowledge Organiser**. These are clear, easy-to-understand, concise summaries of the content that you need to know for your exam. The information is organised to show how one idea flows into the next so you can learn how all the science is tied together, rather than lots of disconnected facts.

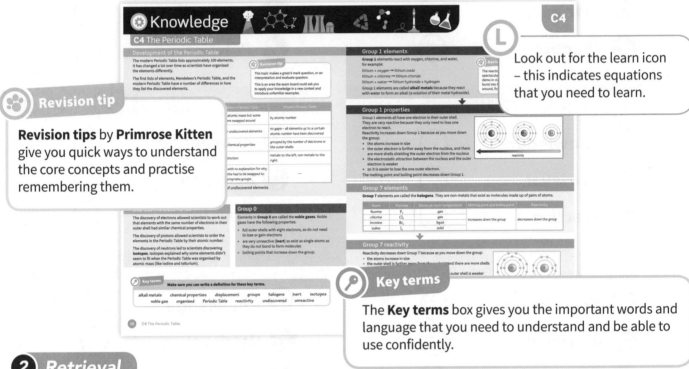

Look out for the learn icon – this indicates equations that you need to learn.

Revision tips by **Primrose Kitten** give you quick ways to understand the core concepts and practise remembering them.

Key terms

The **Key terms** box gives you the important words and language that you need to understand and be able to use confidently.

2 Retrieval

The **Retrieval questions** help you learn and quickly recall the information you've acquired. These are short questions and answers about the content in the Knowledge Organiser. Cover up the answers with some paper; write down as many answers as you can from memory. Check back to the Knowledge Organiser for any you got wrong, then cover the answers and attempt *all* the questions again until you can answer all the questions correctly.

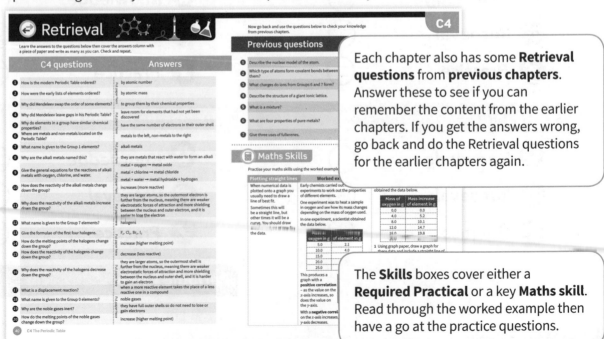

Each chapter also has some **Retrieval questions** from **previous chapters**. Answer these to see if you can remember the content from the earlier chapters. If you get the answers wrong, go back and do the Retrieval questions for the earlier chapters again.

The **Skills** boxes cover either a **Required Practical** or a key **Maths skill**. Read through the worked example then have a go at the practice questions.

Make sure you revisit the retrieval questions on different days to help them stick in your memory. You need to write down the answers each time, or say them out loud, otherwise it won't work.

③ Practice

Once you think you know the Knowledge Organiser and Retrieval answers really well you can move on to the final stage: **Practice**.

Each chapter has lots of **exam-style questions**, including some questions from previous chapters, to help you apply all the knowledge you have learnt and can retrieve.

Each question has a difficulty icon that shows the level of challenge.

 These questions build your confidence.

 These questions consolidate your knowledge.

 These questions stretch your understanding.

Make sure you attempt all of the questions no matter what grade you are aiming for.

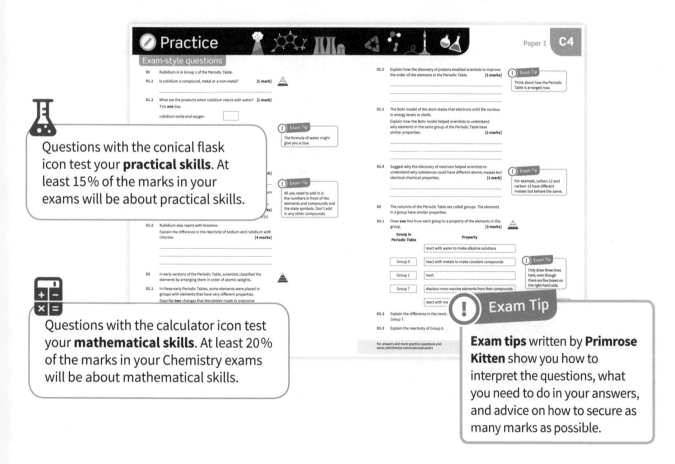

Questions with the conical flask icon test your **practical skills**. At least 15% of the marks in your exams will be about practical skills.

Questions with the calculator icon test your **mathematical skills**. At least 20% of the marks in your Chemistry exams will be about mathematical skills.

Exam tips written by **Primrose Kitten** show you how to interpret the questions, what you need to do in your answers, and advice on how to secure as many marks as possible.

All the **answers** are on Kerboodle and the website, along with even more exam-style questions. www.oxfordrevise.com/scienceanswers

⚙ Knowledge

C1 The atom

Development of the model of the atom

Dalton's model

John Dalton thought of the **atom** as a solid sphere that could not be divided into smaller parts. His model did not include **protons**, **neutrons**, or **electrons**.

The plum pudding model

Scientists' experiments resulted in the discovery of sub-atomic charged particles. The first to be discovered were electrons – tiny, negatively charged particles.

The discovery of electrons led to the plum pudding model of the atom – a cloud of positive charge, with negative electrons embedded in it. Protons and neutrons had not yet been discovered.

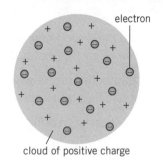
electron
cloud of positive charge

Alpha scattering experiment

1 Scientists fired small, positively charged particles (called alpha particles) at a piece of gold foil only a few atoms thick.
2 They expected the alpha particles to travel straight through the gold.
3 They were surprised that some of the alpha particles bounced back and many were deflected (alpha scattering).
4 To explain why the alpha particles were repelled the scientists suggested that the positive charge and mass of an atom must be concentrated in a small space at its centre. They called this space the **nucleus**.

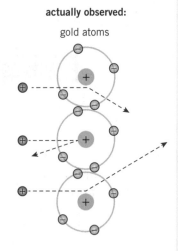
scientists predicted:
gold atoms
actually observed:
gold atoms
alpha particle

Nuclear model

Scientists replaced the plum pudding model with the nuclear model and suggested that the electrons **orbit** (go around) the nucleus, but not at set distances.

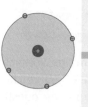

Electron shell (Bohr) model

Niels Bohr calculated that electrons must orbit the nucleus at fixed distances. These orbits are called **shells** or **energy levels**.

The proton

Further experiments provided evidence that the nucleus contained smaller particles called protons. A proton has an opposite charge to an electron.

Size

The atom has a radius of 1×10^{-10} m. Nuclei (plural of nucleus) are around 10 000 times smaller than atoms and have a radius of around 1×10^{-14} m.

Relative mass

One property of protons, neutrons, and electrons is **relative mass** – their masses compared to each other. Protons and neutrons have the same mass, so are given a relative mass of 1. It takes almost 2000 electrons to equal the mass of a single proton – their relative mass is so small that we can consider it as 0.

The neutron

James Chadwick carried out experiments that gave evidence for a particle with no charge. Scientists called this the neutron and concluded that the protons and neutrons are in the nucleus, and the electrons orbit the nucleus in shells.

Atoms and particles

The Periodic Table lists over 100 types of atoms that differ in the number of protons, neutrons, and electrons they each have.

	Relative charge	Relative mass	
Proton	+1	1	= atomic number
Neutron	0	1	= mass number – atomic number
Electron	–1	0 (very small)	= same as the number of protons

All atoms have equal numbers of protons and electrons, meaning they have no overall charge:

total negative charge from electrons = total positive charge from protons

Drawing atoms

Electrons in an atom are placed in fixed shells. You can put

- up to two electrons in the first shell
- eight electrons each in the second and third shells.

You must fill up a shell before moving on to the next one.

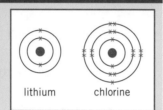

lithium   chlorine

Elements and compounds

Elements are substances made of one type of atom. Each atom of an element will have the same number of protons.

Compounds are made of different types of atoms chemically bonded together. The atoms in a compound have different numbers of protons.

Isotopes

Atoms of the same element can have a different number of neutrons, giving them a different overall mass number. Atoms of the same element with different numbers of neutrons are called **isotopes**.

The **relative atomic mass** is the average mass of all the atoms of an element (note that **abundance** means the percentage of atoms with a certain mass):

$$\text{relative atomic mass} = \frac{(\text{abundance of isotope 1} \times \text{mass of isotope 1}) + (\text{abundance of isotope 2} \times \text{mass of isotope 2})...}{100}$$

Mixtures

- A mixture consists of two or more elements or compounds that are not chemically combined together.
- The substances in a mixture can be separated using physical processes.
- These processes do not use chemical reactions.

Separating mixtures

- filtration – insoluble solids and a liquid
- crystallisation – soluble solid from a solution
- simple distillation – solvent from a solution
- fractional distillation – two liquids with similar boiling points
- paper chromatography – identify substances from a mixture in solution

 **Key terms**

Make sure you can write a definition for these key terms.

abundance	atom	atomic number	compound	electron	element
energy level	isotope	neutron	nucleus	orbit	proton
relative atomic mass	relative charge	relative mass	shell	

Learn the answers to the questions below then cover the answers column with a piece of paper and write down as many as you can. Check and repeat.

C1 questions

Answers

	C1 questions	Answers
1	What is an atom?	smallest part of an element that can exist
2	What is Dalton's model of the atom?	atoms as solid spheres that could not be divided into smaller parts
3	What is the plum pudding model of the atom?	sphere of positive charge with negative electrons embedded in it
4	What did scientists discover in the alpha scattering experiment?	some alpha particles were deflected by the gold foil – this showed that an atom's mass and positive charge must be concentrated in one small space (the nucleus)
5	Describe the nuclear model of the atom.	dense nucleus with electrons orbiting it
6	What did Niels Bohr discover?	electrons orbit in fixed energy levels (shells)
7	What did James Chadwick discover?	uncharged particle called the neutron
8	Where are protons and neutrons?	in the nucleus
9	What is the relative mass of each sub-atomic particle?	proton: 1, neutron: 1, electron: 0 (very small)
10	What is the relative charge of each sub-atomic particle?	proton: +1, neutron: 0, electron: −1
11	How can you find out the number of protons in an atom?	the atomic number on the Periodic Table
12	How can you calculate the number of neutrons in an atom?	mass number – atomic number
13	Why do atoms have no overall charge?	equal numbers of positive protons and negative electrons
14	How many electrons would you place in the first, second, and third shells?	up to 2 in the first shell and up to 8 in the second and third shells
15	What is an element?	substance made of one type of atom
16	What is a compound?	substance made of more than one type of atom chemically joined together
17	What is a mixture?	two or more substances not chemically combined
18	What are isotopes?	atoms of the same element (same number of protons) with different numbers of neutrons
19	What are the four physical processes that can be used to separate mixtures?	filtration, crystallisation, distillation, fractional distillation, chromatography
20	What is relative mass?	the average mass of all the atoms of an element

Put paper here

Required Practical Skills

Practise answering questions on the required practicals using the example below.
You need to be able to apply your skills and knowledge to other practicals too.

Chromatography	Worked Example	Practice

Chromatography

This practical shows the separation of coloured substances by making paper chromatograms. You need to be able to describe the method of chromatography, including the solutes and solvents involved, and define the stationary and mobile phases.

Make sure you know how to calculate R_f values:

$$R_f \text{ value} = \frac{\text{distance moved by solute}}{\text{distance moved by solvent}}$$

Food colourings are often used for this practical, but remember any coloured mixture could be used in an exam question.

Worked Example

A student carried out a paper chromatography experiment to determine what inks make up a sample. They observed the following results.

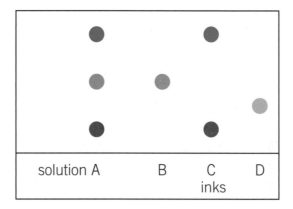

1 Determine which of the inks (B, C, and D) are present in solution A.

Compare the spots on the chromatogram to determine which inks make up solution A – the spots for inks B and C all align with spots for solution A, so these are all present in solution A.

2 Give a reason why an original line is drawn in pencil.

Pencil does not interact with the mobile phase, and therefore will not interfere with the chromatogram.

Practice

1 Two students were setting up a chromatography experiment. Student A wanted to leave the experiment until the solvent front had moved three-quarters of the way up the paper, and student B wanted to leave it for 15 minutes. Which method do you agree with? Give an explanation for your answer.

2 In the experiment in **1**, a spot of ink moved 5 cm and the solvent front moved 17 cm. Calculate the R_f value.

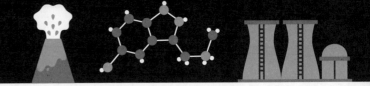
Exam-style questions

01 Different scientists made different contributions to the development of the model of the atom.

01.1 Draw **one** line from each scientist to the contribution they made.   **[2 marks]**

| the nucleus contains neutrons |

| Niels Bohr | | atoms contain electrons |

| James Chadwick | | electrons orbit the nucleus at certain distances |

| | atoms are tiny spheres |

01.2 Describe the alpha particle scattering experiment, and how it showed that the mass of an atom was concentrated in a positively charged nucleus.   **[6 marks]**

_____

_____

_____

_____

_____

_____

_____

_____

> **! Exam Tip**
>
> Ensure you cover how the results showed the presence of a nucleus.

01.3 An atom of potassium has the symbol $^{39}_{19}K$.

What is the atomic number of potassium?

Tick **one** box.   **[1 mark]**

19 ☐

20 ☐

39 ☐

58 ☐

02   A student has a mixture of isopropanol and water. **Table 1** has some physical properties of isopropanol.

Table 1

Melting point of isopropanol	−89 °C
Boiling point of isopropanol	80.3 °C
Solubility of isopropanol in water	very

02.1   Name the method that the student should use to separate the mixture to collect pure samples of each substance.   **[1 mark]**

_____

02.2   Explain how the substances will be separated by the method given in **02.1**.   **[6 marks]**

_____

_____

_____

_____

_____

_____

_____

_____

> **! Exam Tip**
>
> Use data from the table in your answer, and clearly link it to the method of separation.

02.3   Ethanol has a boiling point of 78.4 °C and a melting point of −114.7 °C. Suggest why this method could not be used to separate a mixture of isopropanol and ethanol.   **[1 mark]**

_____

_____

03   Models of the atom have changed over time.

03.1   Compare the plum pudding model of the atom to the earlier model.   **[3 marks]**

03.2   Explain how experimental evidence led scientists to suggest the nuclear model of the atom.   **[6 marks]**

> **! Exam Tip**
>
> You'll need to mention things that are the same and things that are different.

04 **Table 2** gives some information about four different atoms.

The atoms are represented by the letters **W**, **X**, **Y**, and **Z**.

These letters are not the chemical symbols of the elements.

Table 2

Atom	Number of protons	Number of neutrons	Number of electrons
W	16	16	
X	17	20	17
Y	18	22	18
Z	17	18	17

04.1 Give the number of electrons in atom **W**.   **[1 mark]**

04.2 Give the atomic number of atom **X**.   **[1 mark]**

04.3 Give the letter of the atom that has the greatest mass number.   **[1 mark]**

04.4 Give the letters of the **two** atoms that are isotopes of the same element.   **[1 mark]**

> ! **Exam Tip**
>
> The clue in the question is that these are atoms.

05 **Figure 1** shows the electronic structure of an atom. The atom has no overall charge.

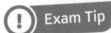

Figure 1

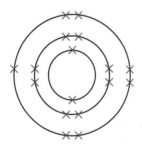

05.1 Identify the number of protons in the nucleus of the atom.   **[1 mark]**

05.2 Give the atomic number of the atom.   **[1 mark]**

05.3 **Figure 1** shows a chlorine atom. Chlorine has a relative atomic mass of 35.5. Explain why the relative atomic mass of chlorine is not a whole number.   **[2 marks]**

05.4 Two isotopes of copper are copper-63 and copper-65. **Table 3** shows the abundance of each isotope.

Table 3

Copper isotope	Percentage abundance
copper-63	69.2
copper-65	30.8

Calculate the relative atomic mass of copper. Give your answer to three significant figures.   **[3 marks]**

> ! **Exam Tip**
>
> Remember protons have a positive charge and electrons have a negative charge.

> ! **Exam Tip**
>
> If you're not sure how to answer this then think of having 69.2 atoms that have a mass of 63 and 30.8 atoms that have a mass of 65. Then just work out the average mass of those 100 atoms.

05.5 Suggest why the relative atomic mass of copper on the Periodic Table is different from the value calculated in **05.4**. **[1 mark]**

06 **Figure 2** shows the electronic structure of an atom. The atom has 12 neutrons.

Figure 2

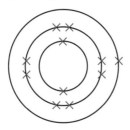

06.1 Deduce the mass number of the atom. Explain how you worked out your answer. **[2 marks]**

06.2 The atom in **Figure 2** has 11 electrons. An atom of argon has 18 electrons.
Compare the arrangements of the electrons in the two atoms. **[6 marks]**

06.3 The radius of an argon atom is 71 pm. Determine the radius of an argon nucleus. Give your answer, in pm, in standard form. **[2 marks]**

07 **Table 4** shows data for three isotopes of an element.
The total percentage abundance of all three isotopes is 100%.

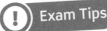

Table 4

Isotope	Number of protons	Number of neutrons	Percentage abundance
L	14	14	92.20
M	14	15	4.68
N	14	16	

07.1 Calculate the relative atomic mass of the element.
Give your answer to three significant figures. **[5 marks]**

07.2 Draw the complete electronic structure for an atom of the element represented in **Table 4**. **[2 marks]**

07.3 Compare the chemical properties of isotope **L** and isotope **N**. **[2 marks]**

08 An atom of silicon has 14 electrons.

08.1 Give the relative charge of an electron. **[1 mark]**

08.2 **Figure 3** shows the energy levels (shells) of the electrons in a silicon atom.

Complete the diagram by drawing the 14 electrons in the correct shells. **[1 mark]**

Figure 3

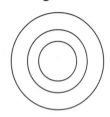

Exam Tip

Start from the centre and work your way out.

08.3 Name the scientist who suggested that electrons orbit the nucleus at specific distances. **[1 mark]**

09 Magnesium exists as three isotopes.

The symbols of the three isotopes are $^{24}_{12}Mg$, $^{25}_{12}Mg$, and $^{26}_{12}Mg$.

09.1 Define the term isotope. **[1 mark]**

09.2 Which statement about the three isotopes of magnesium is true?

Choose **one** answer. **[1 mark]**

The three isotopes have the same mass number.

The three isotopes have the same atomic number.

Each isotope has a different number of electrons.

Each isotope has a different number of protons.

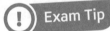

Exam Tip

Start by crossing off the ones you know are incorrect.

09.3 The relative abundances of the three isotopes of magnesium are shown in **Table 5**.

Table 5

Isotope	Percentage abundance
$^{24}_{12}Mg$	79.0
$^{25}_{12}Mg$	10.0
$^{26}_{12}Mg$	11.0

Exam Tip

The percentage abundances add up to 100.

Calculate the relative atomic mass of magnesium. Give your answer to three significant figures. **[4 marks]**

10 **Table 6** gives some information about the most common isotopes of some elements.

Table 6

Element	Number of protons	Number of neutrons
neon	10	10
calcium	20	20
zinc	30	34
zirconium	40	50
tin	50	70
lanthanum	57	82

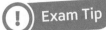

10.1 Write the name of the element that has an atomic number of 40. **[1 mark]**

10.2 Write the name of the element that has a mass number of 40. **[1 mark]**

10.3 Write the electronic structure of a calcium atom. **[1 mark]**

10.4 Plot the data from **Table 6** as a scatter graph on **Figure 4**. **[3 marks]**

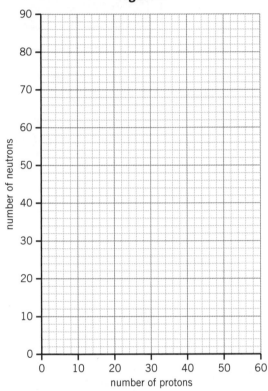
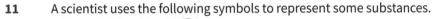

Figure 4

> **! Exam Tip**
>
> Starting from the inside shell, write down the number of electrons in each shell followed by a comma: x,y,z.

> **! Exam Tip**
>
> Use crosses to show where you have plotted you points.

10.5 Draw a curve of best fit on your graph. **[1 mark]**

10.6 Describe the relationship shown on the graph. **[2 marks]**

> **! Exam Tip**
>
> Lines of best fit need to be smooth and continuous.

11 A scientist uses the following symbols to represent some substances.

- potassium atom:
- sodium atom:
- chlorine atom:
- water:

11.1 Identify the symbol that represents a compound. **[1 mark]**

11.2 The scientist uses the symbols to draw a representation of four samples they have (**Figure 5**).

Figure 5

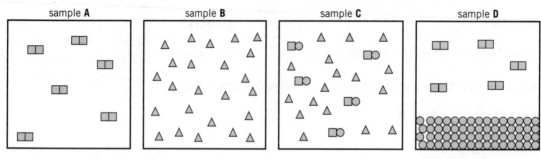

Identify the sample that contains a pure element. **[1 mark]**

11.3 Identify the sample that contains a mixture of two compounds.

[1 mark]

> **! Exam Tip**
>
> Look carefully at the key when answering this question.

11.4 Identify the sample that contains a mixture of two elements.

[1 mark]

11.5 The scientists uses their symbols to draw the following representation of the compound sodium chloride.

Write the chemical formula for this substance. **[1 mark]**

12 **Table 7** shows some atomic radius values.

Table 7

Atom	Atomic radius in nm
carbon	0.077
silicon	0.111
germanium	0.122

The unit nm is a nanometre.

$1\,nm = 1 \times 10^{-9}\,m$

> **! Exam Tip**
>
> You may not have seen nanometres written as a number before – it is very small but don't let that put you off!

12.1 Suggest why silicon atoms are bigger than carbon atoms. **[1 mark]**

12.2 Determine the atomic radius of germanium in metres.

Write your answer in standard form. **[2 marks]**

12.3 Estimate the ratio of the radius of a silicon atom to the radius of a carbon atom.

Give your answer as whole numbers. **[3 marks]**

> **! Exam Tip**
>
> Ratio is a skill you may be used to in maths, but it can come up anywhere in the science course!

12.4 The radius of the nucleus of a carbon atom is $2.7 \times 10^{-15}\,m$.

How many times greater is the radius of a carbon atom than the radius of a carbon nucleus? **[2 marks]**

13 A student has a selection of mixtures:

- sodium chloride salt dissolved in water
- sand and water
- green ink.

13.1 Name the physical process that can be used to separate the sand and water mixture. **[1 mark]**

13.2 Describe a method that the student could use to obtain a sample of pure, dry sodium chloride from the mixture. In your method include any equipment you would use. **[6 marks]**

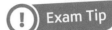

Exam Tip

The key is in the question.

13.3 The student thinks that the green ink is made up of a mixture two dyes.

Name the physical process the student could use to identify whether they are correct. **[1 mark]**

14 A student has an ink. The ink is made of dyes. The dyes are all mixtures. The student used chromatography to try to identify what dyes the ink is made of.

A chromatogram of the ink and three dyes is shown in **Figure 6.**

Figure 6

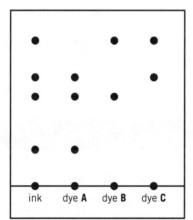

ink   dye **A**   dye **B**   dye **C**

14.1 Identify **one** dye that the student's ink definitely contains. **[1 mark]**

14.2 Explain why the student cannot identify all the dyes contained within the ink from their chromatogram. **[2 marks]**

14.3 Suggest how the student could identify which dye was in the ink. **[1 mark]**

Exam Tip

This is a required practical that you'll cover in more depth later. Use your analysis skills to interpret this set of results.

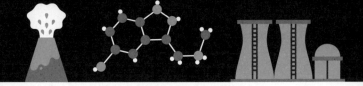

C2 Covalent bonding

Particle model

The three states of matter can be represented in the particle model.

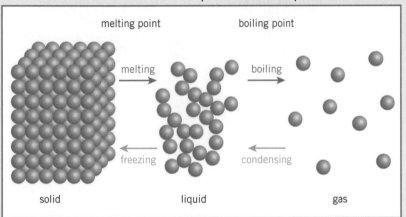

This model assumes that:

- there are no forces between the particles
- that all particles in a substance are spherical
- that the spheres are solid.

The amount of energy needed to change the state of a substance depends on the forces between the particles. The stronger the forces between the particles, the higher the melting or boiling point of the substance.

Covalent bonding

Atoms can share or transfer electrons to form strong chemical bonds.

A **covalent bond** is when electrons are *shared* between **non-metal** atoms.

The number of electrons shared depends on how many extra electrons an atom needs to make a full outer shell.

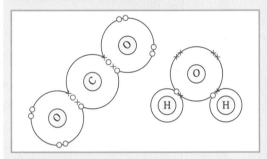

If you include electrons that are shared between atoms, each atom has a full outer shell.
Single bond = each atom shares one pair of electrons.
Double bond = each atom shares two pairs of electrons.

Covalent structures

When atoms form covalent bonds different types of structures can be formed. The structure depends on how many atoms there are and how they are bonded. There are three main types of covalent structure:

Structure and bonding

Giant covalent

Many billions of atoms, each one with a strong covalent bond to a number of others.

An example of a giant covalent structure is diamond.

Small molecules

Each molecule contains only a few atoms with strong covalent bonds between these atoms. Different molecules are held together by weak **intermolecular forces**.

For example, water is made of small molecules.

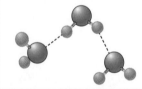

Large molecules

Many repeating units joined by covalent bonds to form a chain.

The small section is bonded to many identical sections to the left and right. The 'n' represents a large number.

Separate chains are held together by intermolecular forces that are stronger than in small molecules.

Polymers are examples of long molecules.

Key terms

Make sure you can write a definition for these key terms.

boiling point   covalent bond   delocalised electrons   double bond   fullerene   giant covalent
graphene   graphite   intermolecular forces   large molecules   melting point   nanotube

Properties	High melting and boiling points because the strong covalent bonds between the atoms must be broken to melt or boil the substances. This requires a lot of energy. Solid at room temperature.	Low melting and boiling points because only the intermolecular forces need to be overcome to melt or boil the substances, not the bonds between the atoms. This does not require a lot of energy as the intermolecular forces are weak. Normally gaseous or liquid at room temperature.	Melting and boiling points are low compared to giant covalent substances but higher than for small molecules. Large molecules have stronger intermolecular forces than small molecules, which require more energy to overcome. Normally solid at room temperature.

Most covalent structures do not conduct electricity because they do not have **delocalised electrons** or ions that are free to move to carry charge.

Graphite

Graphite is a giant covalent structure, but is different to other giant covalent substances.

Structure

Made only of carbon – each carbon atom bonds to three others, and forms hexagonal rings in layers. Each carbon atom has one spare electron, which is delocalised and therefore free to move around the structure.

Hardness

The layers can slide over each other because they are not covalently bonded. Graphite is therefore softer than diamond, even though both are made only of carbon, as each atom in diamond has four strong covalent bonds.

Conductivity

The delocalised electrons are free to move through graphite, so can carry charges and allow an electrical current to flow. Graphite is therefore a conductor of electricity.

Graphene

Graphene consists of only a single layer of graphite. Its strong covalent bonds make it a strong material that can also conduct electricity. It could be used in composites and high-tech electronics.

Fullerenes

- hollow cages of carbon atoms bonded together in one molecule
- can be arranged as a sphere or a tube (called a **nanotube**)
- molecules held together by weak intermolecular forces, so can slide over each other
- conduct electricity

Spheres

Buckminsterfullerene was the first fullerene to be discovered, and has 60 carbon atoms.

Other fullerenes exist with different numbers of carbon atoms arranged in rings that form hollow shapes.

Fullerenes like this can be used as lubricants and in drug delivery.

Nanotubes

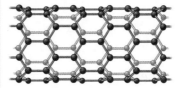

The carbon atoms in nanotubes are arranged in cylindrical tubes.
Their high **tensile strength** (they are difficult to break when pulled) makes them useful in electronics.

non-metal   polymers   single bond   small molecules   tensile strength

⇄ Retrieval

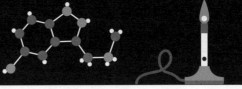

Learn the answers to the questions below then cover the answers column with
a piece of paper and write down as many as you can. Check and repeat.

	C2 questions		Answers
1	How are covalent bonds formed?	Put paper here	by atoms sharing electrons
2	Which type of atoms form covalent bonds between them?	Put paper here	non-metals
3	Describe the structure and bonding of a giant covalent substance.	Put paper here	billions of atoms bonded together by strong covalent bonds
4	Describe the structure and bonding of small molecules.	Put paper here	small numbers of atoms group together into molecules with strong covalent bonds between the atoms and weak intermolecular forces between the molecules
5	Describe the structure and bonding of polymers.	Put paper here	many identical molecules joined together by strong covalent bonds in a long chain, with weak intermolecular forces between the chains
6	Why do giant covalent substances have high melting points?	Put paper here	it takes a lot of energy to break the strong covalent bonds between the atoms
7	Why do small molecules have low melting points?	Put paper here	only a small amount of energy is needed to break the weak intermolecular forces
8	Why do large molecules have higher melting and boiling points than small molecules?	Put paper here	the intermolecular forces are stronger in large molecules
9	Why do most covalent substances not conduct electricity?	Put paper here	do not have delocalised electrons or ions
10	Describe the structure and bonding in graphite.	Put paper here	each carbon atom is bonded to three others in hexagonal rings arranged in layers – it has delocalised electrons and weak forces between the layers
11	Why can graphite conduct electricity?	Put paper here	the delocalised electrons can move through the graphite
12	Explain why graphite is soft.	Put paper here	layers are not bonded so can slide over each other
13	What is graphene?	Put paper here	one layer of graphite
14	Give two properties of graphene.	Put paper here	strong, conducts electricity
15	What is a fullerene?	Put paper here	hollow cage of carbon atoms arranged as a sphere or a tube
16	What is a nanotube?	Put paper here	hollow cylinder of carbon atoms
17	Give two properties of nanotubes.	Put paper here	high tensile strength, conduct electricity
18	Give three uses of fullerenes.	Put paper here	lubricants, drug delivery (spheres), high-tech electronics

Now go back and use the questions below to check your knowledge from previous chapters.

Previous questions | Answers

	Previous questions	Answers
1	What is the relative mass of each sub-atomic particle?	proton: 1, neutron: 1, electron: 0 (very small)
2	What did scientists discover in the alpha scattering experiment?	some alpha particles were deflected by the gold foil – this showed that an atom's mass and positive charge must be concentrated in one small space (the nucleus)
3	What are the four physical processes that can be used to separate mixtures?	filtration, crystallisation, distillation, fractional distillation, chromatography
4	What is a compound?	substance made of more than one type of atom chemically joined together
5	What is the plum pudding model of the atom?	sphere of positive charge with negative electrons embedded in it
6	What is the relative charge of each sub-atomic particle?	proton: +1, neutron: 0, electron: −1
7	What did Niels Bohr discover?	electrons orbit in fixed energy levels (shells)

Put paper here

Put paper here

Put paper here

Maths Skills

Practise your maths skills using the worked example and practice questions below.

Unit conversion

Scientists use different units depending on what is most useful to them. For example, when talking about the size of molecules it doesn't make sense to talk about them in kilometres, so we can use nanometres instead.

Whenever we do a calculation we need to make sure the units are the same, so have to do a unit conversion.

The table below shows you how some units can be compared to each other.

Unit	Standard form in m
1 metre (m)	1×10^{0} m
1 centimetre (cm)	1×10^{-2} m
1 millimetre (mm)	1×10^{-3} m
1 micrometre (μm)	1×10^{-6} m
1 nanometre (nm)	1×10^{-9} m
1 picometre (pm)	1×10^{-12} m

Worked Example

Express 120 cm in metres.

When converting to a larger unit, multiply the original value by the value in metres in standard form.

$120\times1\times10^{-2} = 1.2$ m

Express 120 m in centimetres.

When converting to a smaller unit, divide the original value by the value in metres in standard form.

$= \dfrac{120}{1\times10^{-2}} = 12\,000$ cm

Practice

1 Express 400 cm in metres.

2 Express 20 m in millimetres.

3 Express 0.8 m in nanometres.

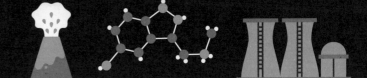

Exam-style questions

01   Silicon dioxide has a giant covalent structure. It has a high melting point and does not conduct electricity.

01.1   Draw **one** line from each property to the explanation of the property.   **[2 marks]**

Exam Tip

Don't be tempted to draw four lines just because there are four boxes on the right.

Only draw **two** lines in total, one from each of the boxes on the left.

Property	Explanation
	strong intermolecular forces of attraction
high melting point	there are no charged particles free to move
does not conduct electricity	strong covalent bonds
	electrons are free to move

01.2   **Figure 1** shows three suggested structures of silicon dioxide, SiO_2.

Figure 1

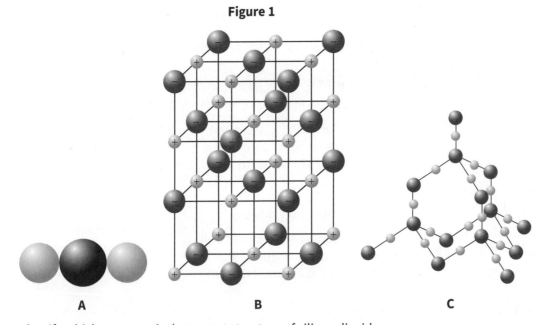

A   B   C

Identify which structure is the correct structure of silicon dioxide.
                                                        [1 mark]

_____

01.3   Silicon dioxide contains the elements silicon and oxygen.
Table 1 shows some properties of silicon and oxygen.

Table 1

	Melting point in °C	Boiling point in °C	Conducts electricity
oxygen	−218.8	−183	no
silicon	1414	3265	yes

Use **Table 1** and the Periodic Table to identify the type of structure of oxygen and silicon. **[2 marks]**

oxygen gas _____

silicon _____

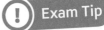

02 Phosphorus is a Group 5 element. It reacts with hydrogen to produce a compound called phosphine.

02.1 The electronic structure of phosphorus is 2,8,5.

Complete the dot and cross diagram in **Figure 2** to show the covalent bonding in a molecule of phosphine, PH_3.

You should show only the electrons in the outer shells. **[2 marks]**

Figure 2

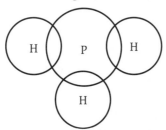

02.2 Name the type of bond or force overcome when liquid phosphine boils. **[1 mark]**

_____

02.3 **Table 2** shows the boiling points of the compounds formed between hydrogen and the elements of Group 5 of the Periodic Table.

Table 2

Formula of compound	Radius of central atom in nm	Boiling point in °C
PH_3	0.110	−88
AsH_3	0.121	−55
SbH_3	0.141	−17
BiH_3	0.152	16

Identify the state that phosphine is in at room temperature. **[1 mark]**

_____

02.4 Describe the trend shown in **Table 2**.
Suggest a reason for this trend. **[2 marks]**

Exam Tip

Increase or decrease isn't enough – an increase or decrease in **what**?

_____

_____

_____

03 Graphene is a single layer of graphite. It can be represented using a ball and stick model.

03.1 The ball and stick model is not a true representation of the structure of graphene. Give **one** reason why. **[1 mark]**

Exam Tip

Don't be worried by very small or very big – first check you can put them into your calculator correctly, then carry out the calculations.

03.2 Explain why graphene conducts electricity. **[1 mark]**

03.3 Graphene is made up of carbon atoms only. One carbon atom has a mass of 1.99×10^{-23} g. A scientist has a sheet of graphene of mass 0.240 g. Calculate the number of carbon atoms in the sheet of graphene. Give your answer to three significant figures. **[3 marks]**

04 Compare the physical properties of diamond and graphite. In your answer, use ideas about bonding to explain the differences in properties. **[6 marks]**

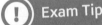

Exam Tip

Ensure you link each property to the feature of it's bonding that gives rise to that property.

05 **Table 3** shows some properties of three elements, **X**, **Y**, and **Z**. The letters are _not_ the chemical symbols of the elements.

Table 3

Element	Melting point in °C	Does the element conduct electricity?
X	−219	no
Y	−101	no
Z	very high	yes

05.1 Identify which element in **Table 3** could be carbon, in the form of graphite. **[1 mark]**

05.2 One of the elements in **Table 3** is chlorine. Draw a dot and cross diagram to show the covalent bonding in a molecule of chlorine, Cl_2. You should show only the electrons in the outer shells. **[2 marks]**

05.3 One of the elements in **Table 3** is oxygen. Draw a dot and cross diagram to show the covalent bonding in a molecule of oxygen, O_2. You should show only the electrons in the outer shells. **[2 marks]**

Exam Tip

The bonding in oxygen gas is a little bit more complicated than in chlorine gas.

05.4 An oxygen atom is smaller than a chlorine atom. Deduce the letter of the element in **Table 3** that represents chlorine. Justify your choice. **[2 marks]**

06 **Figure 3** shows the ball and stick model of a compound, **X**.

Figure 3

06.1 Predict **two** physical properties of compound **X**.

Explain why compound **X** has each of the properties you predicted. **[4 marks]**

06.2 In **Figure 3**, the different coloured balls represent atoms of different elements:

black = carbon   white = hydrogen   red = oxygen.

Deduce the molecular formula of compound **X**. **[1 mark]**

06.3 A sample of compound **X** contains 6.02×10^{23} molecules. Calculate the number of hydrogen atoms in the sample. **[1 mark]**

06.4 The boiling point of compound **X** is 78 °C and the melting point is −114 °C.

Identify the state of compound **X** at 25 °C.

> **! Exam Tip**
>
> Tick of the balls in the diagram once you have counted them, to make sure you don't count some twice.

07 A student had samples of three substances, **X**, **Y**, and **Z**. The student tested which of the substances conduct electricity.

Their results are shown in **Table 4**.

Table 4

Substance	Does it conduct electricity?
X	no
Y	no
Z	yes

07.1 Give the letter of the substance in **Table 4** that could consist of nanotubes. **[1 mark]**

07.2 Give **one** other property of nanotubes. **[1 mark]**

07.3 Give **two** uses of nanotubes. For each use, explain how the properties of nanotubes make them suitable for this use. **[2 marks]**

08 Hydrocarbons are compounds that are made up of carbon atoms and hydrogen atoms only. **Table 5** gives some data on two hydrocarbons.

Table 5

Name of compound	Ball and stick model of molecule	Melting point in °C	Boiling point in °C
methane		−182	−162
hexane		−96	69

In the ball and stick models:

- dark grey spheres represent carbon atoms
- white spheres represent hydrogen atoms

08.1 Write the molecular formula of hexane. **[1 mark]**

08.2 Draw a dot and cross diagram to show the covalent bonding in a molecule of methane, CH_4. You should show only the electrons in the outer shells. **[2 marks]**

08.3 Draw the displayed formula of methane. In the formula, represent each atom with its chemical symbol and each single covalent bond with a line. **[1 mark]**

08.4 Use your own knowledge *and* the data in **Table 5** to compare the physical properties of methane and hexane at room temperature, 20 °C. **[6 marks]**

08.5 Explain, in terms of the forces between molecules, why hexane has a higher boiling point that methane. **[2 marks]**

09 **Table 6** shows the formulae and boiling points of three similar compounds.

Table 6

Name	Formula	Boiling point in °C
fluoroethene	H, H / C=C / F, H	−72
tetrafluoroethene	F, F / C=C / F, F	+76
chloroethene	H, H / C—C / H, Cl	−13
bromoethene	Br, H / C=C / H, H	+16

Exam Tip

You might not be used to seeing compounds drawn like this, but in reality large organic molecules are rarely sitting around in neat straight lines.

Exam Tip

If you're not sure, start by drawing a stick diagram, then a diagram with five overlapping circles, and then add the electrons.

Exam Tip

You may not have seen organic compounds with Halogens in them before – don't panic, just treat them the same as you would any other compound and apply what you know to the question.

09.1 "The more energy levels (shells) of electrons an atom has, the stronger are the intermolecular forces between molecules that contain that atom."

Evaluate this statement using data from **Table 6**. **[6 marks]**

Exam Tip

You *must* include data from the table in your answer and back up each point with an example.

09.2 **Figure 4** shows the formula for a polymer.

Figure 4

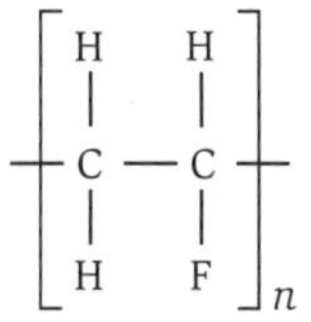

Deduce which compound in **Table 5** is used to make the polymer shown in **Figure 4**. **[1 mark]**

09.3 **Figure 5** shows the formula of the molecules used to make another polymer.

Figure 5

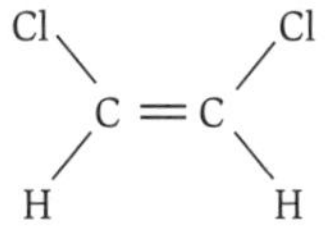

Draw the formula of the polymer in the same form as the polymer formula in **09.2**. **[1 mark]**

Exam Tip

Use your knowledge from class and question **09.2** as examples to base your answer on.

09.4 Explain **two** differences in the physical properties of bromoethene and the polymer that is made from it, poly(bromoethene).

Assume that both substances are at room temperature. **[4 marks]**

10 **Table 7** shows some properties of two oxides.

Table 7

Compound	Formula	Boiling or subliming temperature in °C
carbon dioxide	CO_2	sublimes at −79
silicon dioxide	SiO_2	boils at 2230

10.1 Draw a dot and cross diagram for carbon dioxide. You should show only the electrons in the outer shells. **[2 marks]**

10.2 Explain the difference in the boiling and sublimation temperatures shown in **Table 7**. **[3 marks]**

Exam Tip

Sublimation is when a compounds turns from a solid to a gas, without becoming a liquid.

11   **Table 8** shows the strengths of the covalent bonds in six molecules.

Table 8

Element	Formula of molecule	Bond strength in $kJ\,mol^{-1}$
nitrogen	N_2	944
oxygen	O_2	496
hydrogen	H_2	436
chlorine	Cl_2	242
bromine	Br_2	193
iodine	I_2	151

11.1   Suggest a reason for the difference in bond strengths for Cl_2, Br_2, and I_2.

Use the Periodic Table to help you answer this question.   **[2 marks]**

11.2   Explain the difference in bond strengths for N_2, O_2, and H_2.

Include dot and cross diagrams in your answer.   **[4 marks]**

12.1   Use the data below to estimate the density of the nucleus of the lithium atom, $^{7}_{3}Li$.

Assume that the nucleus is spherical.

Give your answer in $kg\,m^{-3}$ and to one significant figure.

mass of a proton = $1.7\times10^{-27}\,kg$

mass of a neutron = $1.7\times10^{-27}\,kg$

atomic radius of a lithium nucleus = $1\times10^{-14}\,m$

$$density = \frac{mass}{volume}$$

volume of a sphere = $\frac{4}{3}\pi r^3$ (where r = radius)   **[4 marks]**

! Exam Tip

You may not have used this equation in chemistry before, but all you need to do is plug the numbers in.

12.2   In **12.1** you assumed that a lithium nucleus is spherical.

Give **one** assumption made.   **[1 mark]**

12.3   Is this assumption valid?

Explain your answer.   **[2 marks]**

12.4   **Figure 6** shows a simple model of lithium.

Figure 6

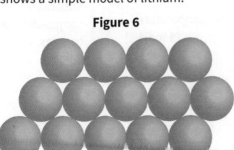

This model also assumes that lithium atoms are spherical.

Give **one** other assumption of this simple model of lithium and explain why the assumption is a limitation of the model.   **[2 marks]**

13 **Table 9** gives the numbers of protons, neutrons, and electrons for some atoms and ions. The atoms and ions are represented by the letters **A** to **E**.

These are not their chemical symbols. You will need to refer to the Periodic Table.

Table 9

Atom, isotope, or ion	Number of protons	Number of neutrons	Number of electrons
A	7	7	7
B	11	12	10
C	12	13	12
D	12	12	10
E	7	8	7

13.1 Write the chemical symbol of **A**, including its mass number, atomic number, and any charge. **[1 mark]**

13.2 Give the letter of the isotope of **A** that is shown in **Table 9**.

Write its chemical symbol, its mass number, its atomic number, and any charge. **[2 marks]**

13.3 Give the letter of the isotope of an ion of **C** that is shown in **Table 8**. Write the chemical formula of the ion of **C**, including its charge. **[2 marks]**

13.4 Give the chemical symbol of the atom in **Table 9** that has the greatest mass number. Write its mass number and atomic number. **[2 marks]**

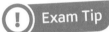

> (!) **Exam Tip**
>
> The atomic number is the clue to the chemical symbol.

14 A student carried out an experiment between solid magnesium and hydrochloric acid. At the end of the experiment, some of the magnesium has not reacted. Describe how the student can separate the unreacted magnesium from the solution. Your description should include the names of any equipment used, where the substances end up, and any relevant safety precautions. **[6 marks]**

> (!) **Exam Tip**
>
> There are some easy marks to be picked up here. For example, there is an obvious safety precaution that will get you some marks.

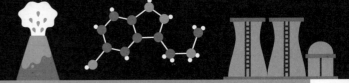

C3 Ionic bonding, metallic bonding, and structure

Ions

As well as sharing electrons, atoms can gain or lose electrons to give them a full outer shell. The number of protons is then different from the number of electrons. The resulting particle has a charge and is called an **ion**.

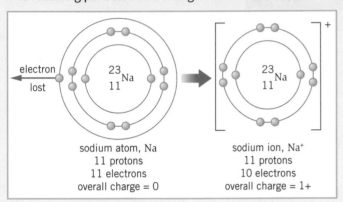

electron lost

sodium atom, Na
11 protons
11 electrons
overall charge = 0

sodium ion, Na⁺
11 protons
10 electrons
overall charge = 1+

Ionic bonding

When metal atoms react with non-metal atoms they **transfer** electrons to the non-metal atom (instead of sharing them).

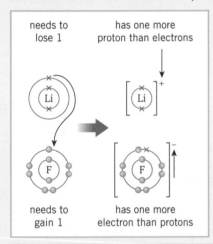

needs to lose 1

has one more proton than electrons

needs to gain 1

has one more electron than protons

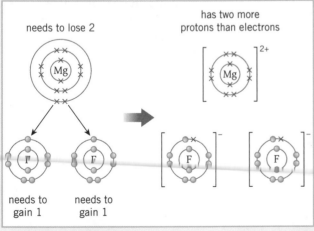

needs to lose 2

has two more protons than electrons

needs to gain 1

needs to gain 1

Metal atoms lose electrons to become positive ions. Non-metal atoms gain electrons to become negative ions.

Ionic structure

Giant ionic lattice

When metal atoms transfer electrons to non-metal atoms you end up with positive and negative ions. These are attracted to each other by the strong **electrostatic force of attraction**. This is called ionic bonding.

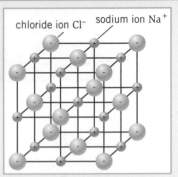

chloride ion Cl⁻   sodium ion Na⁺

The electrostatic force of attraction works in all directions, so many billions of ions can be bonded together in a 3D structure.

Formulae

The formula of an ionic substance can be worked out

1 from its bonding diagram:
   for every one magnesium ion there are two fluoride ions – so the formula for magnesium fluoride is MgF_2

2 from a lattice diagram:
   there are nine Fe^{2+} ions and 18 S^- ions – simplifying this ratio gives a formula of FeS_2

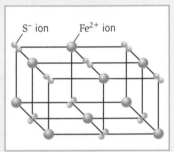

S⁻ ion   Fe²⁺ ion

Ionic properties

Melting points

Ionic substances have high melting points because the electrostatic force of attraction between oppositely charged ions is strong and so requires lots of energy to break.

Conductivity

Solid ionic substances do not conduct electricity because the ions are fixed in position and not free to carry charge.

When melted or dissolved in water, ionic substances do conduct electricity because the ions are free to move and carry charge.

Metals

The atoms that make up metals form layers. The electrons in the outer shells of the atoms are **delocalised** – this means they are free to move through the whole structure.

The positive metal ions are then attracted to these delocalised electrons by the electrostatic force of attraction.

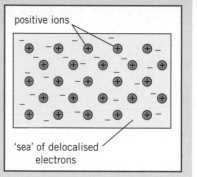

positive ions

'sea' of delocalised electrons

Metallic structure

Pure metals are **malleable** (soft) because the layers can slide over each other.

Metals are good **conductors** of electricity and of thermal energy because delocalised electrons are free to move through the whole structure.

Metals have high melting and boiling points because the electrostatic force of attraction between metal ions and delocalised electrons is strong so lots of energy is needed to break it.

Metallic properties

Alloys

Pure metals are often too soft to use as they are. Adding atoms of a different element can make the resulting mixture harder because the new atoms will be a different size to the pure metal's atoms. This will disturb the regular arrangement of the layers, preventing them from sliding over each other.

The harder mixture is called an **alloy**.

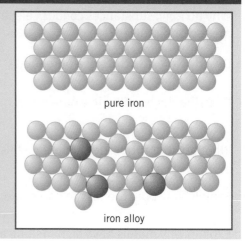

pure iron

iron alloy

Key terms

Make sure you can write a definition for these key terms.

conductivity   conductor   delocalised electron   electrostatic force of attraction
ion   lattice   layer   malleable   transfer

Retrieval

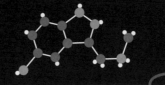

Learn the answers to the questions below then cover the answers column with a piece of paper and write down as many as you can. Check and repeat.

	C3 questions	Answers
1	What is an ion?	atom that has lost or gained electrons
2	Which kinds of elements form ionic bonds?	metals and non-metals
3	What charges do ions from Groups 1 and 2 form?	Group 1 forms 1+, Group 2 forms 2+
4	What charges do ions from Groups 6 and 7 form?	Group 6 forms 2−, Group 7 forms 1−
5	Name the force that holds oppositely charged ions together.	electrostatic force of attraction
6	Describe the structure of a giant ionic lattice.	regular structure of alternating positive and negative ions, held together by the electrostatic force of attraction
7	Why do ionic substances have high melting points?	electrostatic force of attraction between positive and negative ions is strong and requires lots of energy to break
8	Why don't ionic substances conduct electricity when solid?	ions are fixed in position so cannot move, and there are no delocalised electrons
9	When can ionic substances conduct electricity?	when melted or dissolved
10	Why do ionic substances conduct electricity when melted or dissolved?	ions are free to move and carry charge
11	Describe the structure of a pure metal.	layers of positive metal ions surrounded by delocalised electrons
12	Describe the bonding in a pure metal.	strong electrostatic forces of attraction between metal ions and delocalised electrons
13	What are four properties of pure metals?	malleable, high melting/boiling points, good conductors of electricity, good conductors of thermal energy
14	Explain why pure metals are malleable.	layers can slide over each other easily
15	Explain why metals have high melting and boiling points.	electrostatic force of attraction between positive metal ions and delocalised electrons is strong and requires a lot of energy to break
16	Why are metals good conductors of electricity and of thermal energy?	delocalised electrons are free to move through the metal
17	What is an alloy?	mixture of a metal with atoms of another element
18	Explain why alloys are harder than pure metals.	different sized atoms disturb the layers, preventing them from sliding over each other

Put paper here

Previous questions

Answers

	Previous questions		Answers
1	Describe the structure and bonding of a giant covalent substance.		billions of atoms bonded together by strong covalent bonds
2	Why do atoms have no overall charge?		equal numbers of positive protons and negative electrons
3	Why can graphite conduct electricity?		the delocalised electrons can move through the graphite
4	Why do large molecules have higher melting and boiling points than small molecules?		the intermolecular forces are stronger in large molecules
5	What did James Chadwick discover?		uncharged particle called the neutron
6	Give three uses of fullerenes.		lubricants, drug delivery (spheres), high-tech electronics
7	Give two properties of nanotubes.		high tensile strength, conduct electricity
8	How many electrons would you place in the first, second, and third shells?		up to 2 in the first shell and up to 8 in the second and third shells

(Put paper here)

 ## Maths Skills

Practise your maths skills using the worked example and practice questions below.

2D and 3D models	Worked Example	Practice
Scientists often use models to describe what things look like and how they act. These models can be 2D or 3D but they are always just approximations – they are there to help you understand but have strengths and weaknesses.	The model shows how the layers in a metal alloy are disturbed. What are the strengths and weaknesses of this model? 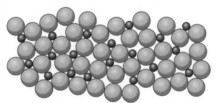 The model is in two dimensions, which helps you to see how the layers are disturbed by atoms of different sizes. However, the metal is normally three dimensional, which this model does not show, so it is not an accurate representation of the metal's structure.	1 Compare and contrast the two models below showing the structure of methane. 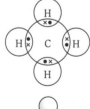

 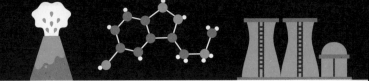
01.1 Which property is typical of metals?
Tick **one** box. **[1 mark]**

They are poor conductors of electricity. ☐

They conduct electricity in the solid state but not in the liquid state. ☐

They conduct electricity in the liquid state but not in the solid state. ☐

They conduct electricity in the solid and liquid states. ☐

! Exam Tip

Only tick one box.
Ticking more than one will mean no marks, even if one of the boxes you've ticked is correct.

01.2 Describe the structure of a pure metal. **[3 marks]**

_____

_____

_____

_____

01.3 Explain why the bonding in a pure metal means that metals can be shaped. **[2 marks]**

_____

_____

_____

! Exam Tip

This is another way of asking why pure metals are soft.

01.4 Mercury is a metal. It is a liquid at room temperature.
Suggest why mercury is an unusual metal. **[1 mark]**

_____

_____

02 Platinum is a metal. It can be used to make jewellery.
Figure 1 shows the arrangement of particles in platinum.

Figure 1

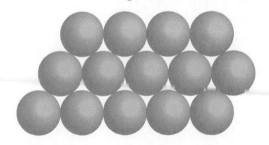

02.1   Explain why platinum can be bent and shaped.

Give your answer in terms of the arrangement of particles in the metal.   **[2 marks]**

_____

_____

_____

02.2   Explain why platinum has a high melting point.

Give your answer in terms of the bonding in the metal.   **[3 marks]**

_____

_____

_____

_____

02.3   Pure platinum is quite soft.

An alloy of platinum that contains rhodium is harder than pure platinum.

Suggest an advantage of making jewellery from a platinum alloy instead of from pure platinum.   **[1 mark]**

_____

_____

02.4   Explain why the platinum alloy is harder than pure platinum.   **[3 marks]**

_____

_____

_____

_____

03 **Figure 2** shows the boiling points of three substances; ethanol (C_2H_5OH), hexanol ($C_6H_{13}OH$), and mercury. Each substance is represented by a letter. The letters are not the chemical symbols of the substances.

Figure 2

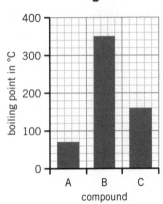

03.1 Give the boiling point of substance **B**. [1 mark]

03.2 Which substance has a boiling point of 78 °C? [1 mark]

03.3 Suggest which letter represents each substance. Explain your answer. [3 marks]

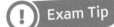

! **Exam Tip**

Think about the structure of each of the compounds and how they will affect their boiling points.

04 Magnesium reacts with oxygen to form magnesium oxide.

04.1 **Figure 3** shows a model of magnesium atoms. Complete **Figure 3** to show the metallic bonding in magnesium. [2 marks]

Figure 3

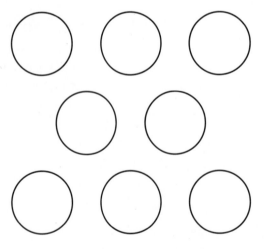

! **Exam Tip**

You'll need to add on positive and negative charges.

04.2 Draw a dot and cross diagram of magnesium oxide. [2 marks]

04.3 Draw **one** line from each substance to the correct property and explanation. **[2 marks]**

Substance	Property and explanation
	conducts electricity in the solid and liquid states because its electrons are free to move
magnesium oxide	conducts electricity in the liquid state only because its electrons are then free to move
magnesium	conducts electricity in the solid and liquid states because its ions are free to move
	conducts electricity in the liquid state only because its ions are then free to move

Exam Tips

Read the instruction in the question carefully.

Not all of the boxes on the right hand side will have lines going into them.

04.4 Write a balanced symbol equation for the reaction between magnesium and oxygen. Include state symbols. **[3 marks]**

04.5 Magnesium and aluminium are two metals. They both have a low density, which makes them lightweight metals. Bicycle wheels can be made of magnesium and aluminium alloys. Explain why an alloy of magnesium and aluminium is used instead of the pure metals.

[2 marks]

Exam Tips

When writing balanced equations:

Step 1: recall the formulae

Step 2: write down the reactants

Step 3: work out the ions

Step 4: determine the formulae of any products

Step 5: check the equation is balanced

Step 6: add state symbols

05 **Table 1** shows the relative conductivities of some metals. The higher the relative conductivity value, the better the metal conducts electricity.

Table 1

Metal	Relative conductivity
aluminium	0.382
beryllium	0.250
lithium	0.108
magnesium	0.224
sodium	0.218
zinc	0.167

05.1 Describe the bonding in pure metals. **[2 marks]**

05.2 "The conductivity of a metal depends on the number of delocalised electrons per atom and which period the metal is in."

Zinc has two electrons in its outer shell. Evaluate the statement above using **Table 1** and the Periodic Table. **[6 marks]**

Exam Tip

You'll have to give evidence for and against this statement and then state a conclusion and justify it with data from the table.

06 **Table 2** shows the melting points of some ionic compounds.

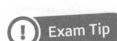

Table 2

Compound	Melting point in °C
calcium oxide	2572
calcium sulfide	2525
calcium bromide	730
magnesium oxide	2852
magnesium sulfide	2000
magnesium bromide	711
sodium oxide	1132
sodium sulfate	884
sodium bromide	747

> **(!) Exam Tip**
>
> It might help to note the charges of each ion next to the table so that it is easy to compare the charges to the melting points.

06.1 Describe the pattern between the charges of the ions in a compound and the melting point of the compound. **[6 marks]**

06.2 Explain the general pattern observed in **Table 2**.

[2 marks]

07 **Figure 4** shows the outer electrons in an atom of magnesium and in an atom of bromine.

Magnesium is in Group 2 of the Periodic Table and bromine is in Group 7.

Figure 4

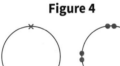

magnesium   bromine

07.1 Magnesium and bromine form an ionic compound. Describe what happens to the electrons when magnesium reacts with bromine. **[3 marks]**

07.2 Give the formulae of the ions formed in the reaction between magnesium and bromine. **[2 marks]**

07.3 Give the chemical formula of the compound formed. **[1 mark]**

07.4 Predict the physical properties of the compound formed. **[3 marks]**

> **(!) Exam Tip**
>
> Bromine only needs one more electron to get a full outer shell and magnesium has two electrons to give away. Think carefully about the ratio of bromine to magnesium.

08   **Table 3** shows the properties of five substances. Each substance is represented by a letter. The letters are not the chemical formulae of the substances.

Table 3

Substance	Melting point in °C	Does it conduct electricity in the solid state?	Does it conduct electricity in the liquid state?
A	993	no	yes
B	1085	yes	yes
C	1263	no	yes
D	1064	yes	yes
E	30	yes	yes

08.1   Use data from **Table 3** to deduce whether substance **C** is a metal or an ionic compound. Justify your answer.   **[3 marks]**

08.2   Give the letters of **two** substances in **Table 3** that could be the element copper. Justify your answer.   **[2 marks]**

08.3   Substance **E** represents the metal gallium. Explain why the melting point of gallium is unusual compared to most other metals.

   [3 marks]

> ⚠ **Exam Tip**
>
> Before you start, mark which compound is ionic and which is covalent, this will prevent you from getting confused later.

09   This question is about two compounds: caesium oxide, Cs_2O, dichlorine monoxide, Cl_2O.

09.1   Draw a dot and cross diagram for dichlorine monoxide.   **[2 marks]**

09.2   Describe the difference in bonding between the two compounds.   **[5 marks]**

09.3   Compare a physical properties of caesium oxide and dichlorine monoxide. Explain your predictions.   **[4 marks]**

09.4   The melting point of caesium oxide is 490 °C. The melting point of barium oxide is 1923 °C. Barium is in Group 2 of the Periodic Table but the same period as caesium.

Suggest why barium oxide has a significantly higher melting point than caesium oxide.   **[2 marks]**

> ⚠ **Exam Tip**
>
> Even if these compounds are unfamiliar to you, the same rules apply as to any other compound. Approach the questions logically and you'll be fine.

10   **Figure 5** shows the outer electrons of a potassium atom and an oxygen atom.

Figure 5

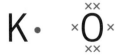

10.1   Draw a dot and cross diagram for the ionic compound formed when oxygen reacts with potassium.   **[2 marks]**

10.2   Describe how the ions are bonded together in potassium oxide.   **[3 marks]**

> ⚠ **Exam Tip**
>
> Remember ionic compounds have square brackets – not overlapping circles.

10.3 The melting point of potassium oxide is 740 °C. The melting point of oxygen is −218 °C.

Explain why the melting point of potassium oxide is much higher than that of oxygen. **[5 marks]**

10.4 The melting point of potassium is 63.5 °C. Give **one** conclusion that can be made about metallic and ionic bonding using this data and the data from **10.3**. **[1 mark]**

11 **Figure 6** shows the structure of part of an ionic compound.

Figure 6

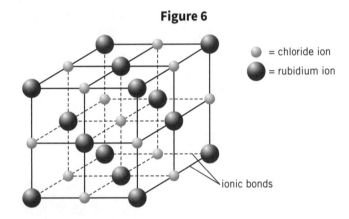

○ = chloride ion

● = rubidium ion

ionic bonds

11.1 Suggest the most likely empirical formula of the ionic compound. **[1 mark]**

11.2 Suggest how **Figure 6** could be improved so that it shows the exact empirical formula of the compound.

Explain your suggestion. **[2 marks]**

11.3 List **two** incorrect assumptions made about the rubidium chloride in the model shown in **Figure 6**. **[2 marks]**

! Exam Tip

There are a large number of atoms shown in the diagram, but this question is asking you for the formula that shows the ratio of atoms in its simplest form.

12 The elements calcium and oxygen react together to form an ionic compound called calcium oxide. Use the Periodic Table to help you answer this question.

12.1 Deduce the charge on a calcium ion and write its formula. **[1 mark]**

12.2 Deduce the charge on an oxide ion and write its formula. **[1 mark]**

12.3 Predict **three** properties of calcium oxide. Explain why calcium oxide has each of these properties. **[6 marks]**

! Exam Tip

Look at the groups that calcium and oxygen are in and determine the number of electrons in their outer shells. This will tell you show many electrons they lose or gain and then you can work out the charge.

13   **Table 4** shows some data about four substances.

Substance	State at 25 °C	Melting point in °C	Boiling point in °C	Conducts electricity when solid	Conducts electricity when liquid
A		−219	−183	no	no
B	solid	1538	2862	yes	yes
C	solid	801	1465	no	yes
D		−7	59	no	no

13.1 Complete **Table 4** to show the state of substance **A** and substance **D**.   **[2 marks]**

13.2 Explain how you can tell that substance **B** is a metal.   **[2 marks]**

13.3 Identify which substance is sodium chloride. Explain your answer.   **[4 marks]**

13.4 Identify which substance in **Table 4** represents oxygen.   **[1 mark]**

13.5 Describe the structure and bonding in substance **D**.   **[2 marks]**

> **! Exam Tip**
> Always use evidence from the data given to you.

14   Copper has two stable isotopes. The chemical symbols of these isotopes are $^{63}_{29}Cu$ and $^{65}_{29}Cu$.

14.1 Give the number of protons in an atom of $^{65}_{29}Cu$.   **[1 mark]**

14.2 Give the number of neutrons in an atom of $^{65}_{29}Cu$.   **[1 mark]**

14.3 Give the mass number of the $^{63}_{29}Cu$ atom.   **[1 mark]**

14.4 Give the number of electrons in a Cu^{2+} ion.   **[1 mark]**

14.5 **Table 5** shows the relative abundances of the two isotopes.

Table 5

Isotope	Percentage abundance
$^{63}_{29}Cu$	69.2
$^{65}_{29}Cu$	30.8

Calculate the relative atomic mass of copper. Give your answer to three significant figures.   **[3 marks]**

> **! Exam Tip**
> This means working out the average mass of 69.2 atoms that have a mass of 63, and 30.8 atoms that have a mass of 65.

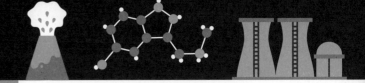

C4 The Periodic Table

Development of the Periodic Table

The modern Periodic Table lists approximately 100 elements. It has changed a lot over time as scientists have organised the elements differently.

The first lists of elements, Mendeleev's Periodic Table, and the modern Periodic Table have a number of differences in how they list the discovered elements.

 **Revision tip**

This topic makes a great 6 mark question, or an interpretation and evaluate question.

This is an area the exam board could ask you to apply your knowledge in a new context and introduce unfamiliar examples.

	First lists of elements	Mendeleev's Periodic Table	Modern Periodic Table
How are elements ordered?	by atomic mass	normally by atomic mass but some elements were swapped around	by atomic number
Are there gaps?	no gaps	gaps left for undiscovered elements	no gaps – all elements up to a certain atomic number have been discovered
How are elements grouped?	not grouped	grouped by chemical properties	grouped by the number of electrons in the outer shells
Metals and non-metals	no clear distinction	no clear distinction	metals to the left, non-metals to the right
Problems	some elements grouped inappropriately	incomplete, with no explanation for why some elements had to be swapped to fit in the appropriate groups	—

Mendeleev was able to accurately predict the properties of undiscovered elements based on the positions of the gaps in his table.

Sub-atomic discoveries

The discovery of electrons allowed scientists to work out that elements with the same number of electrons in their outer shell had similar chemical properties.

The discovery of protons allowed scientists to order the elements in the Periodic Table by their atomic number.

The discovery of neutrons led to scientists discovering **isotopes**. Isotopes explained why some elements didn't seem to fit when the Periodic Table was organised by atomic mass (like iodine and tellurium).

Group 0

Elements in **Group 0** are called the **noble gases**. Noble gases have the following properties:

- full outer shells with eight electrons, so do not need to lose or gain electrons
- are very unreactive (**inert**) so exist as single atoms as they do not bond to form molecules
- boiling points that increase down the group.

 **Key terms**

Make sure you can write a definition for these key terms.

| alkali metals | chemical properties | displacement | groups | halogens | inert | isotopes |
| noble gas | organised | Periodic Table | reactivity | undiscovered | unreactive |

Group 1 elements

Group 1 elements react with oxygen, chlorine, and water, for example:

lithium + oxygen → lithium oxide

lithium + chlorine → lithium chloride

lithium + water → lithium hydroxide + hydrogen

Group 1 elements are called **alkali metals** because they react with water to form an alkali (a solution of their metal hydroxide).

 Revision tip

The reaction of Group 1 metals with water is rather spectacular. You may remember seeing this as a demo in class – some of the tiny lumps of metal burst into flames when they hit the water and whizz around, fizzing as the hydrogen is released.

Group 1 properties

Group 1 elements all have one electron in their outer shell. They are very reactive because they only need to lose one electron to react.

Reactivity increases down Group 1 because as you move down the group:

- the atoms increase in size
- the outer electron is further away from the nucleus, and there are more shells shielding the outer electron from the nucleus
- the electrostatic attraction between the nucleus and the outer electron is weaker
- so it is easier to lose the one outer electron.

The melting point and boiling point decreases down Group 1.

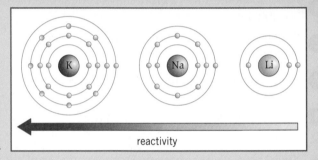

reactivity

Group 7 elements

Group 7 elements are called the **halogens**. They are non-metals that exist as molecules made up of pairs of atoms.

Name	Formula	State at room temperature	Melting point and boiling point	Reactivity
fluorine	F_2	gas		
chlorine	Cl_2	gas	increases down the group	decreases down the group
bromine	Br_2	liquid		
iodine	I_2	solid		

Group 7 reactivity

Reactivity decreases down Group 7 because as you move down the group:

- the atoms increase in size
- the outer shell is further away from the nucleus, and there are more shells between the nucleus and the outer shell
- the electrostatic attraction from the nucleus to the outer shell is weaker
- so it is harder to gain the one electron to fill the outer shell.

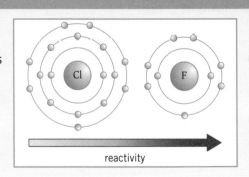

reactivity

Group 7 displacement

More reactive Group 7 elements can take the place of less reactive ones in a compound. This is called **displacement**.

For example, fluorine displaces chlorine as it is more reactive:

fluorine + potassium chloride → potassium fluoride + chlorine

Retrieval
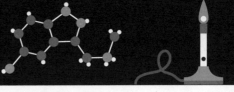

Learn the answers to the questions below then cover the answers column with a piece of paper and write as many as you can. Check and repeat.

	C4 questions		Answers
1	How is the modern Periodic Table ordered?		by atomic number
2	How were the early lists of elements ordered?		by atomic mass
3	Why did Mendeleev swap the order of some elements?		to group them by their chemical properties
4	Why did Mendeleev leave gaps in his Periodic Table?		leave room for elements that had not yet been discovered
5	Why do elements in a group have similar chemical properties?		have the same number of electrons in their outer shell
6	Where are metals and non-metals located on the Periodic Table?		metals to the left, non-metals to the right
7	What name is given to the Group 1 elements?		alkali metals
8	Why are the alkali metals named this?		they are metals that react with water to form an alkali
9	Give the general equations for the reactions of alkali metals with oxygen, chlorine, and water.		metal + oxygen → metal oxide metal + chlorine → metal chloride metal + water → metal hydroxide + hydrogen
10	How does the reactivity of the alkali metals change down the group?		increases (more reactive)
11	Why does the reactivity of the alkali metals increase down the group?		they are larger atoms, so the outermost electron is further from the nucleus, meaning there are weaker electrostatic forces of attraction and more shielding between the nucleus and outer electron, and it is easier to lose the electron
12	What name is given to the Group 7 elements?		halogens
13	Give the formulae of the first four halogens.		F_2, Cl_2, Br_2, I_2
14	How do the melting points of the halogens change down the group?		increase (higher melting point)
15	How does the reactivity of the halogens change down the group?		decrease (less reactive)
16	Why does the reactivity of the halogens decrease down the group?		they are larger atoms, so the outermost shell is further from the nucleus, meaning there are weaker electrostatic forces of attraction and more shielding between the nucleus and outer shell, and it is harder to gain an electron
17	What is a displacement reaction?		when a more reactive element takes the place of a less reactive one in a compound
18	What name is given to the Group 0 elements?		noble gases
19	Why are the noble gases inert?		they have full outer shells so do not need to lose or gain electrons
20	How do the melting points of the noble gases change down the group?		increase (higher melting point)

Put paper here

Now go back and use the questions below to check your knowledge from previous chapters.

C4

Previous questions

Answers

#	Previous questions	Answers
1	Describe the nuclear model of the atom.	dense nucleus with electrons orbiting it
2	Which type of atoms form covalent bonds between them?	non-metals
3	What charges do ions from Groups 6 and 7 form?	Group 6 forms 2−, Group 7 forms 1−
4	Describe the structure of a giant ionic lattice.	regular structure of alternating positive and negative ions, held together by the electrostatic force of attraction
5	What is a mixture?	two or more substances not chemically combined
6	What are four properties of pure metals?	malleable, high melting/boiling points, good conductors of electricity, good conductors of thermal energy
7	Give three uses of fullerenes.	lubricants, drug delivery (spheres), high-tech electronics

Put paper here *Put paper here*

Maths Skills

Practise your maths skills using the worked example and practice questions below.

Plotting straight lines

When numerical data is plotted onto a graph you usually need to draw a line of best fit.

Sometimes this will be a straight line, but other times it will be a curve. You should draw whichever type of line fits the data.

Worked example

Early chemists carried out many experiments to work out the properties of different elements.

One experiment was to heat a sample in oxygen and see how its mass changes depending on the mass of oxygen used.

In one experiment, a scientist obtained the data below.

Mass of oxygen in g	Mass increase of element in g
5.0	2.1
10.0	4.0
15.0	6.2
20.0	8.1
25.0	9.8

This produces a graph with a **positive correlation** – as the value on the x-axis increases, so does the value on the y-axis.

positive correlation +1

With a **negative correlation** the value on the x-axis increases, the value on the y-axis decreases.

Practice

In another experiment, scientists obtained the data below.

Mass of oxygen in g	Mass increase of element in g
0.0	0.0
4.0	5.2
8.0	10.1
12.0	14.7
16.0	19.8
20.0	25.1

1 Using graph paper, draw a graph for these data and include a straight line of best fit.

2 Does your graph show a positive or negative correlation?

3 In another experiment, scientists looked at how the mass of a 5.0 g element increased as it was heated.

Where does the line of best fit start on this graph, compared to on your graph?

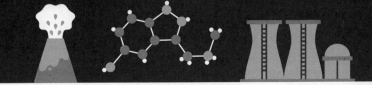

Practice

Exam-style questions

01   Rubidium is in Group 1 of the Periodic Table.

01.1   Is rubidium a compound, metal or a non-metal?   **[1 mark]**

_____

01.2   What are the products when rubidium reacts with water?   **[1 mark]**
Tick **one** box.

rubidium oxide and oxygen   ☐

rubidium hydroxide and oxygen   ☐

rubidium chloride and hydrogen   ☐

rubidium hydroxide and hydrogen   ☐

> **!** **Exam Tip**
>
> The formula of water might give you a clue.

01.3   Rubidium also reacts with oxygen.
Write the word equation for this reaction.   **[1 mark]**

_____

> **!** **Exam Tip**
>
> All you need to add in is the numbers in front of the elements and compounds and the state symbols. Don't add in any other compounds.

01.4   Sodium is another element in Group 1 of the Periodic Table. Sodium reacts with bromine.
Complete the balanced symbol equation for the reaction between sodium and bromine.   **[2 marks]**

_____ Na(_____) + Br$_2$(_____) → _____ NaBr(s)

01.5   Rubidium also reacts with bromine.
Explain the difference in the reactivity of sodium and rubidium with chlorine.   **[4 marks]**

_____

_____

_____

02   In early versions of the Periodic Table, scientists classified the elements by arranging them in order of atomic weights.

02.1   In these early Periodic Tables, some elements were placed in groups with elements that have very different properties.
Describe **two** changes that Mendeleev made to overcome this problem.   **[2 marks]**

_____

_____

Paper 1 C4

02.2 Explain how the discovery of protons enabled scientists to improve the order of the elements in the Periodic Table. **[2 marks]**

_____

_____

_____

! Exam Tip

Think about how the Periodic Table is arranged now.

02.3 The Bohr model of the atom states that electrons orbit the nucleus in energy levels or shells.

Explain how the Bohr model helped scientists to understand why elements in the same group of the Periodic Table have similar properties. **[2 marks]**

_____

_____

_____

02.4 Suggest why the discovery of neutrons helped scientists to understand why substances could have different atomic masses but identical chemical properties. **[1 mark]**

_____

_____

! Exam Tip

For example, carbon-12 and carbon-13 have different masses but behave the same.

03 The columns of the Periodic Table are called groups. The elements in a group have similar properties.

03.1 Draw **one** line from each group to a property of the elements in this group. **[3 marks]**

Group in Periodic Table	Property
	react with water to make alkaline solutions
Group 0	react with metals to make covalent compounds
Group 1	inert
Group 7	displace more reactive elements from their compounds
	react with metals to make ionic compounds

! Exam Tip

Only draw three lines here, even though there are five boxes on the right-hand side.

03.2 Explain the difference in the trend in reactivity down Group 1 and Group 7. **[6 marks]**

03.3 Explain the reactivity of Group 0. **[2 marks]**

! Exam Tip

The answer to **03.2** needs to have two sections.

Make it clear which group you are talking about in each section.

04   Xenon is in Group 0 and Period 5 of the Periodic Table.
Under extreme conditions, xenon will react with fluorine.

04.1   Explain why xenon and fluorine are able to react.   **[3 marks]**

04.2   When xenon reacts with fluorine, the xenon atom
is able to have 12 electrons in its outer shell.
Complete the dot and cross diagram in **Figure 1** to
show the produce of the reaction between xenon
and fluorine.   **[2 marks]**

Figure 1

04.3   Identify the type of bonding in xenon tetrafluroide.   **[1 mark]**

05   A teacher demonstrated the reaction of sodium with water.
This is the method used:

1   Fill a big glass trough with water.

2   Use tongs to take a lump of sodium out of the oil in its
storage bottle.

3   Cut off a small piece of sodium.

4   Put the bigger lump of sodium back in its storage bottle.

5   Use tongs to place the small piece of sodium on the surface of
the water.

05.1   Which group of the Periodic Table is sodium in? Choose **one** answer.
Group 1   Group 2   Group 3   Group 4   **[1 mark]**

05.2   Suggest a reason for step **4**.   **[1 mark]**

05.3   Suggest a reason for using tongs in step **5**.   **[1 mark]**

05.4   In step **5**, the sodium does not start reacting with the water
immediately. Suggest an improvement to step **3** to make the
reaction start more quickly.   **[1 mark]**

05.5   Name the gas made in the reaction.   **[1 mark]**

05.6   Describe how the teacher could show that one of the products of
the reaction makes an alkaline solution in water.   **[2 marks]**

05.7   Explain why the reaction of lithium with water is less vigorous than
the reaction of sodium with water. In your answer, include the
electronic structures of lithium and sodium.   **[3 marks]**

05.8   Caesium is an element near the bottom of Group 1. Predict the
observations on adding caesium to water.   **[1 mark]**

! Exam Tip

Being able to read a method
and suggest improvements is
an important skill in science.

! Exam Tip

First decide if caesium is more
or less reactive than sodium.

06   A student carried out some reactions of halogens with solutions
of potassium chloride, potassium bromide, and potassium iodide.
The solutions were labelled **X**, **Y**, and **Z**. **Table 1** shows the student's
results.

Table 1

Reacted with	Solution X	Solution Y	Solution Z
chlorine water	yellow solution formed	no change observed	brown solution formed
bromine water	no change observed	no change observed	brown solution formed
iodine water	no change observed	no change observed	no change observed

Deduce the identities of solutions **X**, **Y**, and **Z**. Justify your decisions. Use electronic structures to suggest an explanation for one of the reactions that occurs. **[6 marks]**

07 Four pairs of substances are reacted together:

A lithium and bromine   **C** sodium and bromine

B lithium and fluorine   **D** sodium and fluorine

Predict which pair of substances has the most vigorous reaction. Explain your prediction. **[6 marks]**

> **! Exam Tip**
>
> For this answer you'll have to refer to the locations of the elements on the Periodic Table and their structures.

08 **Figure 2** shows the electronic structures of some Group 0 elements. Each is labelled with a letter. The letters are not the chemical symbols of the elements.

08.1 Give the name used for the elements in Group 0 of the Periodic Table. **[1 mark]**

08.2 Which letter represents a helium atom? **[1 mark]**

08.3 Give the letter of the atom of the element in **Figure 2** that has the lowest boiling point. **[1 mark]**

08.4 Draw the electronic structure of a neon atom. **[1 mark]**

08.5 Explain why Group 0 elements do not readily form molecules. **[2 marks]**

Figure 2

> **! Exam Tip**
>
> You get no marks for perfect circles, so try not to spend a long time drawing.

09 **Figure 3** shows the electronic structures of four atoms. Each atom is labelled with a letter. The letters are not the chemical symbols of the elements.

Figure 3

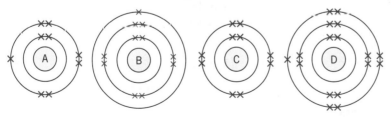

> **! Exam Tip**
>
> Looking at the electrons in the outer shells of these elements is the key to answering the questions.

09.1 Give the letter of the atom of a Group 1 element. **[1 mark]**

09.2 Give the letter of the atom of an unreactive element and explain why this element is unreactive. **[2 marks]**

09.3 Give the letters of two atoms of elements that are in the same group of the Periodic Table. **[1 mark]**

10 **Figure 4** shows the electronic structures of the atoms of three Group 2 elements.

Figure 4

 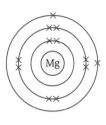 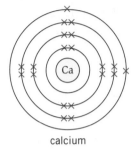

berylium   magnesium   calcium

10.1 Predict how the reactivity of the Group 2 elements changes from the top to the bottom of the group. Justify your prediction by comparing the Group 2 electronic structures to the electronic structures of Group 1. **[4 marks]**

10.2 Describe the structure of calcium. **[3 marks]**

10.3 Magnesium reacts with steam, but does not react with cold water. Is magnesium more or less reactive than sodium? Give a reason for your answer. **[2 marks]**

10.4 Magnesium reacts with chlorine in a similar way to sodium. Give the chemical formula of the product formed when magnesium reacts with chlorine. **[1 mark]**

(!) **Exam Tip**

In written text you can use a shorter notation to refer to the electronic structures. For example, Be would be 2,2, Mg would be 2,8,2, and Ca would be 2,8,8,2.

(!) **Exam Tip**

You'll need to know the formula for ions of magnesium and chlorine to work out the answer.

You can either learn these or work them out from the Periodic Table.

11 The word equations for three reactions are given below.

Reaction 1   hydrogen + fluorine → hydrogen fluoride

Reaction 2   hydrogen + bromine → hydrogen bromide

Reaction 3   iron + bromine → iron bromide

11.1 Draw the electronic structure of the product of reaction **1**. **[2 marks]**

11.2 Explain why reaction **1** is more vigorous than reaction **2**. In your explanation, include the electronic structures of the halogens involved in the reactions. **[6 marks]**

11.3 Predict whether the product of reaction **2** or reaction **3** melts at the higher temperature. Justify your prediction. **[3 marks]**

12.1 Give the name of Group 1 in the Periodic Table. **[1 mark]**

12.2 Why do elements in the same group of the Periodic Table have similar chemical properties? Choose **one** answer. **[1 mark]**

They have the same number of electrons in the shell nearest the nucleus.

They have the same number of electron shells.

(!) **Exam Tip**

Before you answer look up each of the elements in this question on the Periodic Table.

They have the same number of electrons in the shell furthest from the nucleus.

They have the same number of electrons.

12.3 Are Group 1 elements metals or non-metals? **[1 mark]**

12.4 Lithium is a Group 1 element. Lithium reacts with chlorine, a Group 7 element. The product is lithium chloride. Caesium is another Group 1 element, and bromine is another Group 7 element. Name the product when caesium reacts with bromine. **[1 mark]**

13 A student has a dilute potassium chloride solution.

13.1 Which image shows the correct particle diagram for potassium chloride solution? Choose **one** answer. **[1 mark]**

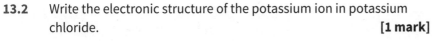

A   B   C   D

13.2 Write the electronic structure of the potassium ion in potassium chloride. **[1 mark]**

13.3 Describe the structure and bonding in potassium chloride. **[4 marks]**

13.4 Give the reason why potassium chloride solution can conduct electricity. **[1 mark]**

13.5 Describe a method by which potassium chloride can be separated from the solution. Your method should result in both potassium chloride and water being collected. **[6 marks]**

14 The Periodic Table lists all of the known elements.

14.1 Explain how the discovery of neutrons led to a greater understanding of the Periodic Table.

14.2 On the Periodic Table, hydrogen is sometimes listed in Group 1 and sometimes listed free above the table. Give **one** reason why hydrogen could be listed in Group 1 and **one** reason why it should not be listed in Group 1. **[2 marks]**

14.3 The modern Periodic Table is arranged by atomic number. What is the atomic number of an element?

14.4 Silicon exists naturally as three isotopes: silicon-28 (92.2%), silicon-29 (4.7%), and silicon-30 (3.1%). Calculate the relative atomic mass of silicon to three significant figures. **[3 marks]**

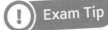

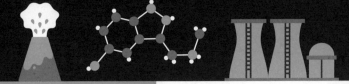

⚙ Knowledge

C5 Transition metals and nanoparticles

Transition Metals

The **Transition Metals** can be found in the middle of the Periodic Table.

Their **properties** are similar to each other, but are different from the Group 1 metals.

Property	Group 1	Transition Metals
melting point	relatively low (e.g., sodium melts at 98 °C)	relatively high (e.g., iron melts at 1538 °C)
density	relatively low	relatively high (e.g., iron is almost ten times as dense as sodium)
strength	relatively low	relatively high
hardness	relatively low (e.g., quite easy to scratch)	relatively high (e.g., quite hard to scratch)
reactivity with oxygen, water, and Halogens	react easily with oxygen, water, and Halogens	react very slowly, if at all, with oxygen, water, and Halogens

Catalysts

Transition Metals are very useful as **catalysts** – they increase the rate of a reaction without being used up.

Coloured compounds

When Transition Metals form compounds, they often take on a colour. For example, chromium(III) oxide is green and cobalt(II) sulfate is red.

Ions

Group 1 metals only form 1+ ions, and Group 2 metals only form 2+ ions. However, most Transition Metals can form many differently charged ions, for example:

- copper can form Cu^+ or Cu^{2+} ions

- manganese can form Mn^{2+}, Mn^{3+}, Mn^{4+}, Mn^{6+}, or Mn^{7+} ions.

In a compound, the **charge** is given by roman numerals in brackets. For example, manganese(II) sulfate contains the Mn^{2+} ion.

Measuring particles

We use different units and scales to measure the size of particles.

Particle	Particulate matter	Size	Standard form	Full form
grain of sand	N/A	0.1 mm	1×10^{-4} m	0.0001 m
coarse particles (e.g., dust)	PM_{10}	10 µm	1×10^{-5} m	0.00001 m
fine particles	$PM_{2.5}$	100 nm	1×10^{-7} m	0.0000001 m
nanoparticles	$< PM_{2.5}$	1 to 100 nm	1×10^{-9} to 1×10^{-7} m	0.000000001 m to 0.0000001 m

PM stands for **particulate matter** and is another way of measuring very small particles.

Surface area-to-volume ratio

Nanoparticles often have very different properties to **bulk** materials of the same substance. This is because of their high surface area-to-volume ratio.

To calculate **surface area-to-volume** ratio follow the steps in the table. As the side length *decreases* by a factor of ten, the surface area-to-volume ratio *increases* by a factor of ten.

	Step	Worked example for cube A	Worked example for cube B
①	work out the surface area of one face of the object	$10\,cm \times 10\,cm = 100\,cm^2$	$1\,cm \times 1\,cm = 1\,cm^2$
②	multiply by number of faces to get total surface area	$100\,cm^2 \times 6 = 600\,cm^2$	$1\,cm^2 \times 6 = 6\,cm^2$
③	calculate volume	$10\,cm \times 10\,cm \times 10\,cm$ $= 1000\,cm^3$	$1\,cm \times 1\,cm \times 1\,cm = 1\,cm^3$
④	write as a ratio	$600:1000$	$6:1$
⑤	divide total surface area (Step 2) by volume (Step 3)	$\dfrac{600\,cm^2}{1000\,cm^3} = 0.6$	$\dfrac{6\,cm^2}{1\,cm^3} = 6$

Uses of nanoparticles

Nanoparticles often have very different properties to bulk materials of the same substance, caused by their high surface area-to-volume-ratio.

Nanoparticles have many uses and are an important area of research. They are used in healthcare, electronics, cosmetics, and as catalysts.

However, nanoparticles have the potential to be hazardous to health and to ecosystems, so it is important that they are researched further.

Key terms

Make sure you can write a definition for these key terms.

bulk   catalyst   charge   density   nanoparticles   particulate matter
properties   surface area-to-volume ratio   Transition Metal

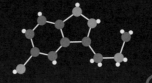

Learn the answers to the questions below then cover the answers column with a piece of paper and write as many as you can. Check and repeat.

C5 questions	Answers
1 Where are Transition Metals found on the Periodic Table?	the middle
2 How do the melting points of Transition Metals compare to those of Group 1 metals?	higher (Transition Metals)
3 How do the densities of Transition Metals compare to those of Group 1 metals?	higher (Transition Metals)
4 How do the strengths of Transition Metals compare to those of Group 1 metals?	higher (Transition Metals)
5 How does the hardness of a Transition Metal compare to that of a Group 1 metal?	higher (Transition Metals)
6 How does the reactivity of Transition Metals with oxygen, Halogens, and water compare to that of Group 1 metals?	lower (Transition Metals)
7 What are Transition Metals used for?	catalysts
8 Give the distinctive properties of Transition Metal compounds.	form ions with different charges and coloured compounds
9 What is the difference between the charges of Transition Metal ions and Group 1 ions?	transition metals can form different charges/Group 1 metals can only form 1+ ions
10 How big are nanoparticles?	1–100 nm
11 How are nanomaterials different from bulk materials?	nanomaterials have a much higher surface area-to-volume ratio
12 What is the relationship between side length and surface area-to-volume ratio?	as side length decreases by a factor of ten, the surface-area-to-volume ratio increases by a factor of ten
13 What are nanoparticles used for?	used in healthcare, electronics, cosmetics, and catalysts
14 How big are fine particles?	100–2500 nm
15 How big are coarse particles?	2.5×10^{-6} to 1×10^{-5} m

Put paper here

Now go back and use the questions below to check your knowledge from previous chapters.

Previous questions / Answers

	Previous questions	Answers
1	Why did Mendeleev leave gaps in his Periodic Table?	leave room for elements that had not yet been discovered
2	Why don't ionic substances conduct electricity when solid?	ions are fixed in position so cannot move, and there are no delocalised electrons
3	How can you find out the number of protons in an atom?	the atomic number on the Periodic Table
4	Why does the reactivity of the halogens decrease down the group?	they are larger atoms, so the outermost shell is further from the nucleus, meaning there are weaker electrostatic forces of attraction and more shielding between the nucleus and outer shell, and it is harder to gain an electron
5	Why do most covalent substances not conduct electricity?	do not have delocalised electrons or ions
6	Give the formulae of the first four halogens.	F_2, Cl_2, Br_2, I_2
7	Give two properties of nanotubes.	high tensile strength, conduct electricity
8	Why are metals good conductors of electricity and of thermal energy?	delocalised electrons are free to move through the metal

Put paper here

Maths Skills

Practise your maths skills using the worked example and practice questions below.

Surface area-to-volume ratio

The steps for working out the surface area to volume ratio of a cube are given in Knowledge Organiser. But you also need to know how to work out the surface areas and volumes of triangles and rectangles.

For a triangle:

$$\text{area} = \text{height} \times \left(\frac{\text{length of base}}{2}\right)$$

For a rectangle:

area = width × length

Units for area are squared.

You may be asked to calculate the volume of a cuboid:

volume = length × height × width

Units for volume are cubed.

Worked Example

1 Calculate the area of a triangle with a base of 5.3 cm and a height of 2.6 cm.

$= \left(\frac{5.3}{2}\right) \times 2.6 = 6.89 = 6.9\,\text{cm}^2$

2 Calculate the area of a rectangle with a width of 0.31 m and a length of 0.18 m.

$= 0.31 \times 0.18 = 0.0558 = 0.056\,\text{m}^2$

3 A cuboid has a height of 12 mm, a length of 15 mm and a width of 7 mm. Calculate its volume.

$= 15 \times 12 \times 7 = 1260\,\text{mm}^3$

Practice

1 Calculate the area of a triangle with a base of 65 cm and a height of 103 cm.

2 Calculate the area of a rectangle with a width of 2.7 km and a length of 0.94 km.

3 A cuboid has a height of 0.48 cm, a length of 0.61 cm, and a width of 0.22 cm. Calculate its volume.

4 Calculate the surface area-to-volume ratio of the cuboid in **3**.

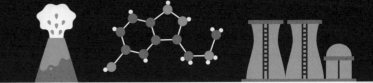
01   Data for two metals are given in **Table 1**.

Each metal is represented by a letter. The letters are **not** the chemical symbols of the metals.

Table 1

Metal	Reaction with oxygen	Colour of its compound with chlorine	Does it conduct electricity?	Observations on placing in water
A	when exposed to air, immediately forms a white coating	white	yes	fizzes, forming bubbles and an alkaline solution
B	small pieces burn in air; reacts very slowly with oxygen in the air at room temperature.	brown	yes	over several days, forms a brown flaky substance

One metal in **Table 1** is a transition metal and one is in Group 1 of the Periodic Table.

01.1   Which physical property in **Table 1 cannot** be used to distinguish between a Group 1 metal and a transition metal?   **[1 mark]**

_____

01.2   Identify which metal from **Table 1** is iron.

Justify your decision.   **[3 marks]**

_____

_____

_____

_____

01.3   Two oxides of iron are FeO and Fe_2O_3.

Explain why the fact that iron forms two oxides is evidence that iron is a transition metal.   **[2 marks]**

_____

_____

_____

02 .1   Draw **one** line from each type of particle to its range of diameters.   **[3 marks]**

Type of particle	Range of diameters

coarse particles

fine particles

nanoparticles

1 to 100 nm

100 to 2500 nm

2.5×10^{-6} m to 1×10^{-5} m

1×10^{-3} m to 2.5×10^{-3} m

Exam Tip

One of the ranges is there to mislead you.

02.2   A gold particle has a diameter of 0.0000034 m. Write this number in standard form.   **[1 mark]**

_____

Exam Tip

You will need to know how to use standard form for your Science exams.

02.3   Is the particle in **02.2** a coarse particle, a fine particle, or a nanoparticle?   **[1 mark]**

_____

02.4   Explain why nanoparticles often have different properties to the same material in bulk.   **[2 marks]**

_____

_____

_____

03   **Figure 1** shows apparatus to measure hardness.

A scientist uses the apparatus in **Figure 1** to compare the hardness of three different metals.

This is the method used:

1   Push the ball down with a force of 30 000 N.

2   Hold for 5 seconds.

3   Remove the ball.

4   Measure the diameter of the indentation on the surface of the metal.

5   Repeat with two different metals.

Figure 1

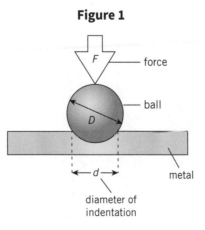

force

ball

D

metal

diameter of indentation

Exam Tip

You might think this experiment belongs in a Physics question; this is just about applying what you know to a new context so don't let that put you off!

03.1 Identify the independent and dependent variables in the experiment. **[2 marks]**

03.2 Explain why the force used is the same for all three metals. **[1 mark]**

03.3 Suggest why the ball in **Figure 1** is made from a very hard material. **[1 mark]**

03.4 Suggest **one** improvement that would reduce the effect of random errors in measuring the diameter of the indentation. **[1 mark]**

03.5 The scientist's results are shown in **Table 2**.

Table 2

Substance	Diameter of indent in mm
A	2
B	6
C	7

[3 marks]

Identify which metal is mostly likely to be the transition metal. Explain your answer.

04 Silver nanoparticles can be used to kill disease-causing bacteria. The nanoparticles enter a bacteria cell through its wall and membrane.

04.1 Suggest **two** reasons that might explain why silver nanoparticles are more effective than bulk silver at killing bacteria. **[2 marks]**

04.2 Suggest **one** possible risk of using nanoparticle silver in medicines for humans. **[1 mark]**

04.3 **Figure 2** shows the percentage of bacteria that survived after coming in contact with different amounts of silver nanoparticles.

Figure 2

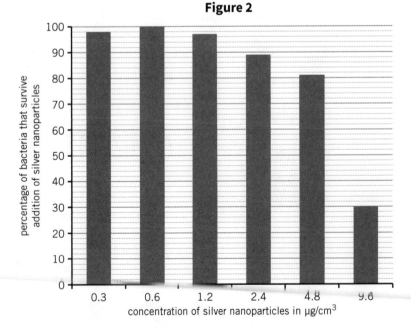

Describe the pattern shown on the bar chart. **[2 marks]**

04.4 Give **two** other uses of nanoparticles. **[2 marks]**

Exam Tip

Go over the method and pick out the independent and dependent variables.

Exam Tip

The first thing to do is to think about the properties of transition metals.

Exam Tip

The main body of the question is short but it provides lots of information to help answer these questions; don't ignore it.

05   **Figure 3** shows the densities of four metals. Each metal is represented by a letter. The letters are **not** the chemical symbols of the metals.

Figure 3

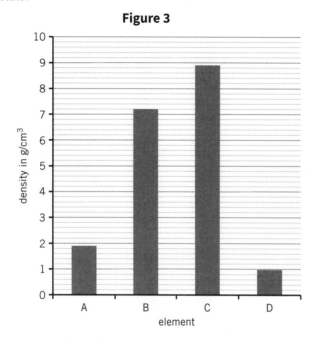

05.1   Give the density of element **C**, to two significant figures. Include units in your answer.   **[2 marks]**

05.2   Give the density of element **B**, to one significant figure.   **[1 mark]**

05.3   Give the letters of the **two** elements that are most likely to be transition metals.   **[1 mark]**

05.4   A student finds the density of chromium. This is the method used:

1   Measure the mass of a lump of chromium.

2   Pour some water into a beaker. Record its volume.

3   Add the chromium to the beaker. Record the new volume, for water plus chromium.

Suggest an improvement to step **2** to measure the water volume more accurately.   **[1 mark]**

05.5   Explain why the method described **cannot** be used to find the density of lithium.   **[1 mark]**

05.6   The student made an improvement to the method. **Table 3** shows the student's results.

Table 3

Mass in g	41
Volume of water at start in cm³	50
Volume of water + chromium in cm³	56

Density is calculated using the equation $\text{density} = \dfrac{\text{mass}}{\text{volume}}$.

Calculate the density of the chromium in g/cm³. Give your answer to two significant figures.   **[3 marks]**

> ! **Exam Tip**
>
> **05.1** asks you to determine a value from the graph to two significant figures. You may not be used to doing this, but it's no different from reading any other type of graph.

> ! **Exam Tip**
>
> Thinking about your Physics knowledge will help you with putting this chemistry in a new context.

06 Use the Periodic Table to answer the following questions.

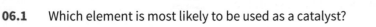

06.1 Which element is most likely to be used as a catalyst? Choose **one** answer. **[1 mark]**

calcium   rhodium   sodium   strontium

06.2 Which element will react with cold water? Choose **one** answer. **[1 mark]**

copper   lithium   titanium   zinc

06.3 Which metal will have the highest density? Choose **one** answer. **[1 mark]**

aluminium   iron   magnesium   potassium

06.4 Which element can form +4 ions. Choose **one** answer. **[1 mark]**

neon   calcium   sodium   vanadium

06.5 A student has a blue metal compound. Which metal is in the compound? Choose **one** answer. **[1 mark]**

copper   lead   rubidium   tin

> **! Exam Tip**
>
> Only select the number of answers you're asked for in multiple choice questions: in this case one. If you circle two answers, you won't get the marks. Equally, don't leave any blank!

07 **Figure 4** is an outline of the Periodic Table.

Each element is represented by a letter. The letters are **not** the chemical symbols of the elements.

Figure 4

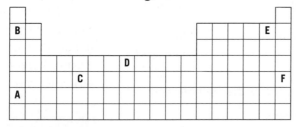

> **! Exam Tip**
>
> Draw a line to divide metals and non-metals. This will help with these questions.

07.1 Give the letters of **two** elements that do not conduct electricity. **[1 mark]**

07.2 Give the letters of the **two** elements that react together most vigorously. **[1 mark]**

07.3 Compare the physical and chemical properties of elements **B** and **D**. **[6 marks]**

Figure 5

08 This question is about nanoparticles. **Figure 5** shows a cubic nanoparticle, 3 nm wide.

08.1 Calculate the surface area to volume ratio of the nanoparticle. **[3 marks]**

> **! Exam Tip**
>
> Remember to think about the sides that you can't see.

08.2 Explain why the properties of a nanoparticle material are different from the properties of the same material in bulk. **[1 mark]**

09 **Table 4** gives the diameters of some particles.

Table 4

Particle	Diameter in nm
gold atom	0.174
water molecule	0.275

09.1 Explain why a water molecule is **not** a nanoparticle. **[1 mark]**

09.2 Write the diameter of a water molecule in metres. Give your answer in standard form. **[2 marks]**

09.3 A certain gold nanoparticle has a cubic shape. The length of a side of the cube is 50 nm. Estimate the number of gold atoms that are on one face of the cube. Give your answer to one significant figure in standard form. **[4 marks]**

10 **Table 5** shows data for some elements.

Table 5

Element	Melting point in °C	Relative conductivity
caesium	29	0.53
copper	1083	5.9
gold	1063	4.2
iron	1535	1.0
lithium	180	1.1
sodium	98	2.2

Transition metals have higher melting points and conductivities than Group 1 elements. Evaluate the statement above using the data in **Table 5** only. **[6 marks]**

Exam Tip

The question says "using the data in **Table 7** only" meaning you won't gain any marks for information you write that isn't in the table or back up by data from the table.

11 Atoms are made up of sub-atomic particles.

11.1 Give the relative charge of each sub-atomic particle. **[3 marks]**

11.2 Which sub-atomic particle determines the identity of an element? **[1 mark]**

11.3 Using the Periodic Table, give the atomic number of oxygen. **[1 mark]**

11.4 Explain why the mass number of chlorine is not a whole number. **[2 marks]**

11.5 Write the electronic structure of a phosphorous atom. **[1 mark]**

12 Sodium reacts with chlorine to form sodium chloride.

12.1 Explain how the structure and bonding in sodium and chlorine give rise to their different properties. **[5 marks]**

12.2 Sodium chloride is an ionic compound. Describe the structure and bonding in sodium chloride. Explain why this compound has a high melting point. **[6 marks]**

Exam Tip

Before you start, determine the charge on the ions of sodium and chlorine.

12.3 Write a balanced symbol equation for the reaction between sodium and chlorine. Include state symbols. **[3 marks]**

⚙ Knowledge

C6 Chemical calculations with mass

Conservation of mass

The conservation of mass states that atoms cannot be created or destroyed in a chemical reaction. Atoms are rearranged into new substances. All the atoms you had in the reactants must be present in the products.

As such, when it comes to measuring the mass of a reaction, you would expect the mass at the start to be the same as the mass at the end. However, sometimes the mass can appear to change.

Decrease in mass

In some reactions the mass appears to decrease. This is normally because a gas is produced in the reaction and lost to the surroundings. For example:

$$sodium + water \rightarrow sodium\ hydroxide + hydrogen$$
$$2Na(s) + 2H_2O(l) \rightarrow 2NaOH(aq) + H_2(g)$$

The mass of the sodium and the water at the start of the reaction will be more than the mass of the sodium hydroxide at the end of the reaction, because hydrogen atoms have been lost as a gas.

Increase in mass

In some reactions the mass appears to increase. This is normally because one of the reactants is a gas. For example:

$$sodium + chlorine \rightarrow sodium\ chloride$$
$$2Na(s) + Cl_2(g) \rightarrow 2NaCl(s)$$

The mass of the sodium at the start of the reaction will be lower than the mass of sodium chloride at the end of the reaction. This is because atoms from the gaseous chlorine have been added to the sodium, increasing the mass.

Balancing symbol equations

When writing symbol equations you need to ensure that the number of each atom on each side is equal.

$$H_2 + O_2 \rightarrow H_2O$$

unbalanced

there are 2 hydrogen atoms on each side, but 2 oxygen atoms in the reactants and 1 in the product

$$2H_2 + O_2 \rightarrow 2H_2O$$

balanced

there are 4 hydrogen atoms on each side, and 2 oxygen atoms on each side

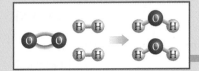

State symbols

A balanced symbol equation should also include state symbols.

State	Symbol
solid	(s)
liquid	(l)
gas	(g)
aqueous or dissolved in water	(aq)

Ratios

Look back at the reaction. In the reaction between hydrogen and oxygen, the ratio of hydrogen to oxygen molecules is 2:1. This means that for every *one* molecule of oxygen, you would need *two* molecules of hydrogen, for example:

- if you had 10 molecules of oxygen you would need 20 molecules of hydrogen
- if you had 1 mole of oxygen you would need 2 moles of hydrogen
- if you had 1.75 moles of oxygen you would need 3.5 moles of hydrogen.

A balanced symbol equation shows the ratios of the reactants and products in a chemical reaction.

Formula mass

Every substance has a **formula mass**, M_r.

formula mass M_r = sum (relative atomic mass of all the atoms in the formula)

Avogadro's constant

One mole of a substance contains 6.02×10^{23} atoms, ions, or molecules. This is **Avogadro's constant**.

One mole of a substance has the same mass as the M_r of the substance. For example, the M_r (H_2O) = 18, so 18 g of water molecules contains $6 \times 6.02 \times 10^{23}$ molecules, and is called one mole of water.

You can write this as: $moles = \dfrac{mass}{M_r}$

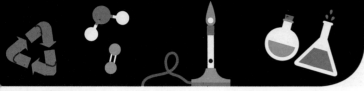

Using balanced equations

In a balanced symbol equation the sum of the M_r of the reactants equals the sum of the M_r of the products.

If you are asked what mass of a product will be formed from a given mass of a specific reactant, you can use the steps below to calculate the result.

1 balance the symbol equation
2 calculate moles of the substance with a known mass using $\text{moles} = \dfrac{\text{mass}}{M_r}$
3 using the balanced symbol equation, work out the number of moles of the unknown substance
4 calculate the mass of the unknown substance using $\text{mass} = \text{moles} \times M_r$

If you are asked to balance an equation, you can use the steps below to work out the answer.

1 work out M_r of all the substances
2 calculate the number of moles of each substance in the reaction using $\text{moles} = \dfrac{\text{mass}}{M_r}$
3 convert to a whole number ratio
4 balance the symbol equation

Excess and limiting reactants

In a chemical reaction between two or more reactants, often one of the reactants will run out before the others. You then have some of the other reactants left over. The reactant that is left over is in **excess**. The reactant that runs out is the **limiting reactant**.

To work out which reactants are in excess and which is the limiting reactant, you need to:

1 write the balanced symbol equation for the reaction
2 pick one of the reactants and its quantity as given in the question
3 use the ratio of the reactants in the balanced equation to see how much of the other reactant you need
4 compare this value to the quantity given in the question
5 determine which reactant is in excess and which is limiting.

Concentration

Concentration is the amount of solute in a volume of solvent. The unit of concentration is g/dm^3. Concentration can be calculated using:

$$\text{concentration } (g/dm^3) = \frac{\text{mass } (g)}{\text{volume } (dm^3)}$$

Sometimes volume is measured in cm^3:

$$\text{volume } (dm^3) = \frac{\text{volume } (cm^3)}{1000}$$

- lots of solute in little solution = high concentration
- little solute in lots of solution = low concentration

 Revision tip

Sometimes you may get concentration in mol/dm^3 and need to convert it into g/dm^3.

concentration in g/dm^3 = concentration in $mol/dm^3 \times Mr$

 **Key terms**

Make sure you can write a definition for these key terms.

Avogadro's constant	balanced	calculation	concentration	conservation
dm^3	equation	excess	formula mass	limiting reactant	mass
	mole	ratio	state	surroundings	

Retrieval

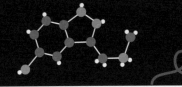

Learn the answers to the questions below then cover the answers column with
a piece of paper and write as many as you can. Check and repeat.

C6 questions	Answers
1 What is the conservation of mass?	in a chemical reaction, atoms are not created or destroyed, just rearranged, so total mass before = total mass after the reaction
2 When a metal forms a metal oxide, why does the mass increase?	atoms from gaseous oxygen have been added
3 When an acid reacts with a metal, why does the mass decrease?	a gas is produced and escapes
4 What is relative formula mass?	the sum of the relative atomic masses of each atom in a substance
5 What are the four state symbols and what do they stand for?	(s) solid, (l) liquid, (g) gas, (aq) aqueous or dissolved in water
6 How can you tell when a symbol equation is balanced?	the number of atoms of each element is the same on both sides
7 What is a mole?	mass of a substance that contains $6.02{\times}10^{23}$ particles
8 Give the value for Avogadro's constant.	$6.02{\times}10^{23}$
9 Which formula is used to calculate the number of moles from mass and M_r?	$\text{moles} = \dfrac{\text{mass}}{M_r}$
10 Which formula is used to calculate the mass of a substance from number of moles and M_r?	$\text{mass} = \text{moles} \times M_r$
11 What is a limiting reactant?	the reactant that is completely used up in a chemical reaction
12 What is a unit for concentration?	g/dm^3 or mol/dm^3
13 Which formula is used to calculate concentration from mass and volume?	$\text{concentration (g/dm}^3\text{)} = \dfrac{\text{mass (g)}}{\text{volume (dm}^3\text{)}}$
14 Which formula is used to calculate volume from concentration and mass?	$\text{volume (dm}^3\text{)} = \dfrac{\text{mass (g)}}{\text{concentration (g/dm}^3\text{)}}$
15 Which formula is used to calculate mass from concentration in g/dm^3 and volume?	$\text{mass (g)} = \text{concentration (g/dm}^3\text{)} \times \text{volume (dm}^3\text{)}$
16 How can you convert a volume reading in cm^3 to dm^3?	divide by 1000
17 If the amount of solute in a solution is increased, what happens to its concentration?	increases
18 If the volume of water in a solution is increased, what happens to its concentration?	decreases

Put paper here

Now go back and use the questions below to check your knowledge from previous chapters.

C6

Previous questions

Answers

	Previous questions		Answers
1	What is the relationship between side length and surface area-to-volume ratio?	Put paper here	as side length decreases by a factor of ten, the surface-area-to-volume ratio increases by a factor of ten
2	What is the difference between the charges of Transition Metal ions and Group 1 ions?		transition metals can form different charges/Group 1 metals can only form 1+ ions
3	How are covalent bonds formed?		by atoms sharing electrons
4	How can you calculate the number of neutrons in an atom?	Put paper here	mass number – atomic number
5	How do the melting points of Transition Metals compare to those of Group 1 metals?		higher (Transition Metals)
6	Explain why alloys are harder than pure metals.	Put paper here	different sized atoms disturb the layers, preventing them from sliding over each other
7	How does the reactivity of Transition Metals with oxygen, Halogens, and water compare to that of Group 1 metals?		lower (Transition Metals)
8	Give the general equations for the reactions of alkali metals with oxygen, chlorine, and water.		metal + oxygen → metal oxide metal + chlorine → metal chloride metal + water → metal hydroxide + hydrogen
9	Give two properties of graphene.		strong, conducts electricity

 Maths Skills

Practise your maths skills using the worked example and practice questions below.

Orders of magnitude	Worked Example	Practice
An order of magnitude is when you look at the difference in values with reference to powers of ten. For example, 200 is larger than 20 by 180, but 200 is *one order of magnitude larger* because 20 × 10 = 200. Similarly, 7000 is three orders of magnitude larger than 7.	A piece of marble has a length of 60 cm. A smaller piece of marble has a length of 0.6 cm. By how many orders of magnitude do they differ in size? **Step 1:** Divide the bigger number by the smaller one: $\frac{60}{0.6} = 100$ **Step 2:** Count the zeros – 100 has two zeros. **Step 3:** The number of zeros = orders of magnitude, so the marbles differ in size by two orders of magnitude.	1 A square has a length of 40 cm. A larger square has a length of 40 000 cm. By how many orders of magnitude is the larger square longer than the smaller one? 2 A solution contains 20 000 hydrogen ions. After a reaction, it contains only 20 hydrogen ions. By how many orders of magnitude does the number of hydrogen ions differ before and after the reaction? 3 Explain how the pH scale uses orders of magnitude.

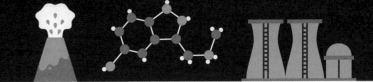

Practice

Exam-style questions

01 Methane is a compound with the formula CH_4.

01.1 Calculate the relative formula mass M_r of methane. **[1 mark]**

Relative atomic masses A_r: H = 1; C = 12

_____

_____

Relative formula mass: _____

> **! Exam Tip**
>
> Show your working clearly.

01.2 Methane reacts with excess oxygen to make carbon dioxide and water.

Methane is the limiting reactant.

What is meant by the term limiting reactant?

Tick **one** box. **[1 mark]**

the reactant present in the smaller mass ☐

the reactant with the smaller relative formula mass ☐

the reactant with the smaller molar mass ☐

the reactant that is completely used up when the other reactant is present in excess ☐

01.3 Write a balanced symbol equation for the reaction between methane and oxygen. **[2 marks]**

_____

01.4 0.13 moles of methane react with 0.25 moles of oxygen. Which reactant is the limiting reactant? **[1 mark]**

_____

01.5 How many moles of water will be produced in the reaction in **01.4**? **[1 mark]**

_____

> **! Exam Tip**
>
> You'll need to use the equation for this. If you didn't get the equation correct you can still get some marks by showing your working.

02   This question is about calcium nitrate, $Ca(NO_3)_2$.

02.1   How many oxygen atoms are there in 1 mole of calcium nitrate?
Avogadro constant = 6.02×10^{23} **[2 marks]**

>  **Exam Tip**
>
> Take careful note of which number is inside the brackets and which is outside the brackets.

_____

02.2   What is the relative formula mass of calcium nitrate? Use the Periodic Table to help you.

Tick **one** box.   **[1 mark]**

102   ☐

150   ☐

164   ☐

204   ☐

02.3   In a fume cupboard, a student heats some calcium nitrate in a test tube.

The calcium nitrate decomposes:

$$2Ca(NO_3)_2(s) \rightarrow 2CaO(s) + 4NO_2(g) + O_2(g)$$

Relative atomic masses A_r: Ca = 40, N = 14, O = 16.

Explain why the mass of solid in the test tube is **lower** after the chemical reaction.   **[1 mark]**

> **!** **Exam Tip**
>
> State symbols are not often used at GCSE, so if you see them in a question there is a good chance you need to refer to them in the answer. Look at the changes of state to find the answer to this one.

_____

_____

02.4   In the reaction, 22.4 g of calcium oxide are produced. Calculate the mass of calcium nitrate that reacted.   **[5 marks]**

_____

_____

_____

_____

mass = _____ g

03   1 cm³ of water is equal to 1 g of water at room temperature and pressure.

03.1   Determine the volume in cm³ of 1 mole of water.   **[2 marks]**

! Exam Tip

Start with working out the M_r of water.

03.2   Calculate the volume in cm³ of one molecule of water.   **[2 marks]**

03.3   Use the particle model to explain why the value calculated in **03.2** is not accurate.   **[2 marks]**

04   Sulfur dioxide reacts with oxygen to make sulfur trioxide.

$$SO_2(g) + 2O_2(g) \rightarrow 2SO_3(g)$$

04.1   Write down what the number 3 means in the formula SO_3.   **[1 mark]**

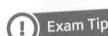

! Exam Tip

Even for a one mark question its important to show your working.

04.2   Calculate the relative formula mass M_r of sulfur dioxide, SO_2. Relative atomic masses A_r: S = 32; O = 16.   **[1 mark]**

04.3   In an experiment, 1.28 g of sulfur dioxide, SO_2, makes 1.68 g of sulfur trioxide, SO_3. Calculate the mass of oxygen that was needed.   **[1 mark]**

05.1   What is the Avogadro constant the measure of?   **[1 mark]**

05.2   A glass contains 232 g of water. Estimate the number of water molecules in the glass. Give your answer to three significant figures. Relative atomic masses A_r: H = 1; O = 16.   **[5 marks]**

05.3   Deduce the number of water molecules in 464 g of ice. Use your answer to **05.2**.   **[2 marks]**

06   A teacher makes sodium chloride by adding burning sodium to a container of chlorine gas.

06.1   Suggest **one** safety precaution the teacher should take.   **[1 mark]**

06.2   Balance the symbol equation for the reaction and add state symbols.   **[2 marks]**

$$\underline{} \ Na + Cl_2 \rightarrow \underline{} \ NaCl$$

06.3   Describe the structure and bonding in solid sodium chloride. In your answer, outline how the ions are made and give their charges.   **[6 marks]**

! Exam Tip

To begin it might help to think about:

- how many electrons sodium has in its outer shell and what happens to them
- the charge on the sodium ions
- how many electrons chlorine has in its outer shell and what happens to them
- the charge on the chloride ions.

07 Some students investigated the reaction of calcium carbonate with hydrochloric acid:

$$CaCO_3(s) + 2HCl(aq) \rightarrow CaCl_2(aq) + CO_2(g) + H_2O(l)$$

The students measured the volume of carbon dioxide gas made in 60 s. The students repeated the experiment five times. **Table 1** shows their results.

Table 1

Experiment number	Volume of carbon dioxide gas made in 60 s in cm³
1	52
2	49
3	48
4	56
5	55

07.1 Calculate the mean volume of carbon dioxide gas. **[1 mark]**

07.2 Give the range of the values obtained in the five experiments. **[1 mark]**

07.3 What is the best estimate of the volume of gas obtained?
Choose **one** answer. **[1 mark]**

mean ± 3   mean ± 4   mean ± 6   mean ± 8

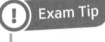
Exam Tip

You might not have seen numbers written like this before. It is just asking for how high above and below the mean the outermost values are.

07.4 Hydrochloric acid solution contains hydrogen chloride, HCl, molecules dissolved in water. The students used 25 cm³ of 7.3 g/dm³ hydrochloric acid. Calculate the mass of hydrogen chloride that dissolved. Give your answer to two significant figures. **[3 marks]**

08 A student wants to react nitric acid with potassium hydroxide to form potassium nitrate and water. The balanced symbol for the equation is:

$$HNO_3 + KOH \rightarrow KNO_3 + H_2O$$

08.1 Complete the symbol equation by adding state symbols. **[1 mark]**

08.2 The student dissolved 14 g of potassium hydroxide in 700 cm³ of water. Calculate the concentration of the potassium hydroxide solution in g/dm³. **[2 marks]**

08.3 The concentration of nitric acid was 22 g/dm³. Calculate the mass of nitric acid in 30 cm³ of the solution. **[2 marks]**

08.4 The student reacted 30 cm³ of nitric acid with 35 cm³ of this solution of potassium hydroxide. Identify the limiting reactant. **[6 marks]**

09 Magnesium reacts with nitrogen gas, N_2, to make magnesium nitride, Mg_3N_2.

Exam Tip

Don't worry about unfamiliar compounds; you do not need to draw magnesium nitride.

09.1 Draw a dot and cross diagram to show the bonding in a nitrogen molecule, N_2. **[2 marks]**

09.2 Two groups of students draw diagrams of apparatus they think could be used to make magnesium nitride from magnesium and nitrogen (**Figure 2**).

Figure 2

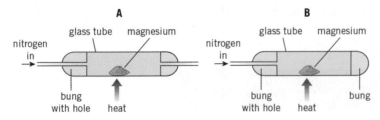

! Exam Tip

Carefully compare the two images to look for the differences.

Explain why the apparatus in **B** must **not** be used for the experiment. **[2 marks]**

09.3 A teacher made magnesium nitride from magnesium and nitrogen. **Table 2** shows the masses of the reactants that reacted and the mass of product made.

Table 2

Substance	Mass in g
magnesium	2.16
nitrogen	0.84
magnesium nitride	3.00

! Exam Tip

You can work backwards from the mass.

Use the data in **Table 2** to deduce the balanced equation for the reaction. Show all your working and use the data below. Relative atomic masses A_r: Mg = 24; N = 14. **[5 marks]**

10 Paracetamol and ibuprofen are painkillers.

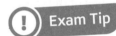

10.1 A solution contains 500 mg of paracetamol in 5 cm³ of solution. Calculate the mass of paracetamol in 1 dm³ of solution. Give your answer in g. **[3 marks]**

10.2 The chemical formula of paracetamol is $C_8H_9NO_2$. Calculate the mass of 1 mole of paracetamol. Relative atomic masses A_r: C = 12; H = 1; N = 14; O = 16. **[1 mark]**

! Exam Tip

There are lots of parts to this question, write everything down clearly to avoid missing or repeating parts.

10.3 A solution of ibuprofen contains 0.10 g of ibuprofen in 5.0 cm³ of solvent. The chemical formula of ibuprofen is $C_{13}H_{18}O_2$. Calculate the number of moles of ibuprofen in 1 dm³ of the solution. Give your answer to three significant figures. **[4 marks]**

11 Some people take iron tablets if they do not have enough iron in their blood. **Table 3** gives some data about three types of iron tablet.

Table 3

Name of compound	Formula of iron compound in tablet	Mass of iron compound in tablet in g
iron(II) sulfate	$FeSO_4$	0.065
iron(II) fumarate	$C_4H_2FeO_4$	0.076
iron(II) gluconate	$C_{12}H_{24}FeO_{14}$	0.300

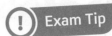

11.1 Calculate the relative formula mass of iron(II) fumarate. Relative atomic masses A_r: Fe = 56; C = 12; H = 1; O = 16. **[1 mark]**

11.2 Calculate the number of moles of iron(II) sulfate in one tablet. Give your answer in standard form to two significant figures. Relative atomic masses A_r: Fe = 56; S = 32; O = 16. **[4 marks]**

11.3 Deduce the mass of iron in one iron(II) gluconate tablet. **[4 marks]**

! **Exam Tip**

Check the answer to **11.2** before you try **11.3**. They are very similar methods and its best to correct any mistakes before you move on.

12 Iron is extracted from its ore in the following reaction. Relative atomic mases A_r: Fe = 56, C = 12, O = 16.

$$2Fe_2O_3 + 3C \rightarrow 4Fe + 3CO_2$$

12.1 Calculate the mass of carbon that reacts with 16.0 g of iron(III) oxide. **[5 marks]**

12.2 Calculate the mass of carbon dioxide produced in the reaction of carbon with 16.0 g of iron(III) oxide. **[4 marks]**

12.3 An industrial plant processes 3.7 tonnes of iron(III) oxide. Calculate the mass in kg of iron produced. 1 tonne = 1000 kg **[6 marks]**

12.4 Calculate the percentage by mass of iron(III) oxide that is iron. **[2 marks]**

13 A teacher demonstrates the reaction of sodium with chlorine. This is the method used:

1 Heat a small piece of sodium.

2 Fill a gas jar with chlorine.

3 Place the hot sodium in the gas jar of chlorine.

13.1 Suggest **two** safety precautions the teacher should take. **[2 marks]**

13.2 Explain an improvement to the order in which the steps above are carried out. **[2 marks]**

13.3 Draw a dot and cross diagram of the product of the reaction. **[2 marks]**

13.4 Describe the bonding in the product. **[3 marks]**

13.5 Explain the difference in the conductivity of electricity between sodium, chlorine, and the product of the reaction. **[6 marks]**

13.6 Predict **one** difference in the observations made if chlorine was replaced by bromine. **[1 mark]**

! **Exam Tip**

Think about the substances being used in the reaction; they might give you a clue about the safety precautions required.

14 Phosphorus has 15 electrons.

14.1 Sketch the electronic structure of phosphorous. **[1 mark]**

14.2 Deduce the group number that phosphorus is in in the Periodic Table. **[1 mark]**

14.3 Compare the properties of protons, neutrons, and electrons. Include in your answer the location of each type of sub-atomic particle within an atom. **[6 marks]**

 # Knowledge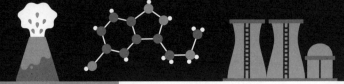

C7 Chemical calculations with moles

Theoretical yield

The **theoretical yield** of a chemical reaction is the mass of a product that you expect to be produced.

Even though no atoms are gained or lost during a chemical reaction, it is not always possible to obtain the theoretical yield because

- some of the product can be lost when it is separated from the reaction mixture
- there can be unexpected side reactions between reactants that produce different products
- the reaction may be reversible.

The theoretical yield of a product can be calculated using the balanced symbol equation, as shown in Chapter 6 *Quantitative chemistry 1*.

Percentage yield

The **yield** is the amount of product that you actually get in a chemical reaction.

Percentage yield is the actual yield as a proportion of the theoretical yield:

$$\text{percentage yield} = \frac{\text{actual yield}}{\text{theoretical yield}} \times 100$$

Atom economy

The **atom economy** of a reaction tells you the proportion of atoms that you started with that are part of *useful* products.

High atom economies are more sustainable, as they mean fewer atoms are being wasted in products that are not useful.

The percentage atom economy is calculated by:

$$\text{atom economy} = \frac{M_r \text{ of useful product}}{M_r \text{ of all products}} \times 100$$

Moles of gases

At any given temperature and pressure, the same number of moles of a gas will occupy the same volume.

At room temperature (25 °C) and pressure (1 atm), one mole of *any* gas will occupy 24 dm³.

To calculate the number of moles of a gas:

$$\text{moles of a gas} = \frac{\text{volume (dm}^3)}{24\,\text{dm}^3}$$
or
$$\text{moles of a gas} = \frac{\text{volume (cm}^3)}{24\,000\,\text{cm}^3}$$

For example, to calculate the number of moles of 3 dm³ of carbon dioxide at room temperature and pressure:

$$\text{number of moles} = \frac{3}{24} = 0.125\,\text{mol}$$

> mol is the unit of moles

0.125 moles of carbon dioxide, oxygen, and hydrogen will all have the volume 3 dm³.

To calculate the volume of a gas at room temperature and pressure:

$$\text{volume (dm}^3) = \text{moles} \times 24\,\text{dm}^3$$

For example, to calculate the volume of 0.25 mol of chlorine at room temperature and pressure:

$$\text{volume (dm}^3) = 0.25\,\text{mol} \times 24\,\text{dm}^3 = 6\,\text{dm}^3$$

 **Key terms**

Make sure you can write a definition for these key terms.

| atom economy | burette | concordant | end point | percentage yield | pipette |
| room temperature and pressure | theoretical yield | titration | titre | useful | yield |

Concentration in mol/dm³

Concentration can also be measured in mol/dm³.

$$\text{concentration of solution (mol/dm}^3) = \frac{\text{number of moles of solute}}{\text{volume of solution (dm}^3)}$$

You can use this formula and mass = moles × M_r to calculate the mass of solute dissolved in a solution.

- The greater the mass of solute in solution, the greater the number of moles of solute, and therefore the greater the concentration.
- If the same number moles of solute is dissolved in a smaller volume of solution, the concentration will be greater.

Titration

Titration is an experimental technique to work out the concentration of an unknown solution in the reaction between an acid and an alkali.

1 Use a pipette to extract a known volume of the solution with an unknown concentration. A pipette measures a fixed volume only.
2 Add the solution of unknown concentration to a conical flask and put the conical flask on a white tile.
3 Add a few drops of a suitable indicator to the conical flask.
4 Add the other solution with a known concentration to the burette.
5 Carry out a rough titration to find out approximately what volume of solution in the burette needs to be added to the solution in the conical flask. Add the solution from the burette to the solution in the conical flask 1 cm³ at a time until the end point is reached.
6 The end point is when the indicator just changes colour.
7 Record the volume of the end point as your rough value.
8 Now repeat steps 1–7, but as you approach the end point add the solution from the burette drop-by-drop. Swirl the conical flask in between drops.
9 Record the volume of the end point.

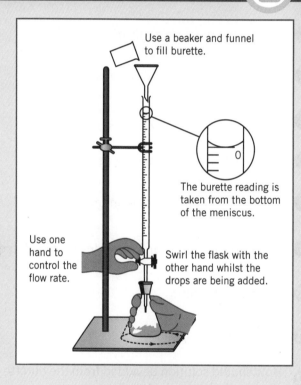

Use a beaker and funnel to fill burette.

The burette reading is taken from the bottom of the meniscus.

Use one hand to control the flow rate.

Swirl the flask with the other hand whilst the drops are being added.

Concordance

Repeat the titration until you get **concordant titres**.

- A titre is the volume of solution that you have added from your burette.
- Concordant means that the titres are within 0.1 cm³ of each other.

You would stop when you had two concordant results, and calculate the mean:

$$\text{mean} = \frac{\text{sum of the concordant results}}{\text{number of concordant results}}$$

Calculating concentration

To calculate the concentration of the unknown solution (the solution in the conical flask):

1 Write a balanced symbol equation for the reaction.
2 Calculate the moles used from the known solution using:

$$\text{moles} = \text{concentration (mol/dm}^3) \times \text{volume (dm}^3)$$

3 Use the ratio from the balanced symbol equation to deduce the number of moles present in the unknown solution.
4 Calculate the concentration of the unknown solution using: $\text{concentration (mol/dm}^3) = \dfrac{\text{moles}}{\text{volume (dm}^3)}$

 # Retrieval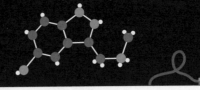

Learn the answers to the questions below then cover the answers column with
a piece of paper and write as many as you can. Check and repeat.

C7 questions	Answers
1 What is the yield of a reaction?	mass of product obtained from the reaction
2 What is the theoretical yield of a reaction?	maximum mass of the product that could have been produced
3 Why is the actual yield always less than the theoretical yield?	• reaction may be reversible • some of the product can be lost on separation • unexpected side reactions between reactants
4 What is the percentage yield?	actual yield as a proportion of theoretical yield
5 How is percentage yield calculated?	$\dfrac{\text{actual yield}}{\text{theoretical yield}} \times 100$
6 What is atom economy?	measure of how many atoms of the reactants end up as useful products
7 Why is a high atom economy desirable?	results in less waste/is more sustainable
8 How is percentage atom economy calculated?	$\dfrac{M_r \text{ of useful product}}{M_r \text{ of all products}} \times 100$
9 How can concentration in mol/dm³ be calculated?	$\dfrac{\text{moles of solute}}{\text{volume (dm}^3\text{)}}$
10 What is a titration?	method used to calculate the concentration of an unknown solution
11 What is the end-point?	the point at which the reaction is complete (when the indicator changes colour) and no substance is in excess
12 How should solution be added from the burette close to the end point?	drop by drop, swirling in between
13 Why is a white tile used in titration?	to see the colour change better
14 What is a titre?	volume of solution added from the burette
15 What are concordant titres?	titres within 0.1 cm³ of each other
16 What volume does one mole of any gas occupy at room temperature and pressure?	24 dm³ or 24 000 cm³

(Put paper here)

 Revision tip

You might be asked to give concentration in g/dm³ instead of mol/dm³.

concentration in g/dm³ = concentration in mol/dm³ × *Mr*

Now go back and use the questions below to check your knowledge from previous chapters.

Previous questions

Answers

	Previous questions	Answers
1	What are the four state symbols and what do they stand for?	(s) solid, (l) liquid, (g) gas, (aq) aqueous or dissolved in water
2	What is relative mass?	the average mass of all the atoms of an element
3	Describe the structure and bonding in graphite.	each carbon atom is bonded to three others in hexagonal rings arranged in layers – it has delocalised electrons and weak forces between the layers
4	Explain why metals have high melting and boiling points.	electrostatic force of attraction between positive metal ions and delocalised electrons is strong and requires a lot of energy to break
5	How are nanomaterials different from bulk materials?	nanomaterials have a much higher surface area-to-volume ratio
6	How can you convert a volume reading in cm^3 to dm^3?	divide by 1000
7	What is a displacement reaction?	when a more reactive element takes the place of a less reactive one in a compound
8	What are nanoparticles used for?	used in healthcare, electronics, cosmetics, and catalysts
9	Give the value for Avogadro's constant.	6.02×10^{23}

(Put paper here)

 # Required Practical Skills

Practise answering questions on the required practicals using the example below. You need to be able to apply your skills and knowledge to other practicals too.

Neutralisation reactions	Worked Example	Practice
You need to be able to use titration to determine the concentration of a solution of acid or alkali. To do this, you need to be able to describe how to carry out a titration experiment using burettes and pipettes, and how to accurately measure and transfer volumes of liquids. You should also be able to use a volume to calculate the concentration of a solution.	A student carried out a titration, adding sulfuric acid to sodium hydroxide. They repeated their experiment until they had carried out four titres. 1 Describe what concordant results are. Concordant results for a titration are results that are within $0.10\,cm^3$ of each other. A titration should be repeated until you have at least three concordant titres. 2 The student got the following results for the volume of sulfuric acid in cm^3 needed to neutralise $25\,cm^3$ sodium hydroxide: 16.05, 16.10, 16.25, 16.00. Calculate the mean volume of acid needed. $16.25\,cm^3$ is not a concordant result so is discarded. $$\text{mean volume} = \frac{(16.05 + 16.10 + 16.00)}{3} = 16.05\,cm^3$$	1 Describe how to take a reading from a burette. 2 Explain how the end of point of a titration can be determined, naming a suitable indicator and colour changes in your answer.

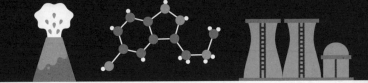

Exam-style questions

01.1 Draw **one** line from each measure of the effectiveness of a chemical reaction to the equation used to calculate it. **[2 marks]**

Measure

Equation

$$\frac{\text{mass of product actually made}}{\text{maximum theoretical mass of product}} \times 100\,\%$$

percentage yield

$$\frac{\text{sum of relative formula masses of all reactants from equation}}{\text{relative formula mass of desired product from equation}} \times 100\,\%$$

atom economy

$$\frac{\text{relative formula mass of desired product from equation}}{\text{sum of relative formula masses of all reactants from equation}} \times 100\,\%$$

$$\frac{\text{maximum theoretical mass of product}}{\text{mass of product actually made}} \times 100\,\%$$

> **! Exam Tip**
>
> In chemistry you need to learn all of the equations, you don't get a formula sheet like you do in physics.

01.2 Give **two** reasons why it is important to use reactions with high atom economy in industry. **[2 marks]**

1 _____

_____

2 _____

_____

01.3 A student makes some lead iodide. They used the following method:

1 Mix a solution of lead nitrate with a solution of potassium iodide. Lead iodide forms as a yellow precipitate.

2 Filter the mixture. Lead iodide is the residue on the filter paper.

3 Use a spatula to scrape the lead iodide from the filter paper.

4 Weigh the lead iodide.

Suggest why the mass of lead iodide is less than the maximum theoretical yield. **[1 mark]**

> **! Exam Tip**
>
> Read over the method carefully to find the answer.

_____

_____

02 A teacher demonstrated the thermite reaction. In the thermite reaction, iron(III) oxide reacts with aluminium. Large amounts of heat are transferred to the surroundings.

The equation for the reaction is:

$$Fe_2O_3 + 2Al \rightarrow 2Fe + Al_2O_3$$

Relative atomic masses A_r: Fe = 56; O = 16; Al = 27

02.1 Suggest **two** safety precautions that the teacher should take.

[2 marks]

1 _____

_____

2 _____

_____

02.2 The teacher used 8.0 g of iron(III) oxide and 2.7 g of aluminium. Show that neither reactant is present in excess. **[4 marks]**

_____
_____
_____
_____

02.3 Calculate the maximum theoretical yield of iron made from 8.0 g of iron(III) oxide. **[2 marks]**

_____ g

02.4 The iron made in the reaction was weighed. Its mass was 4.6 g. Calculate the percentage yield of iron in the reaction. Use your answer to **02.3**. **[2 marks]**

_____ %

02.5 Suggest **two** reasons why the percentage yield of the thermite reaction is not 100 %. **[2 marks]**

1 _____

_____

2 _____

_____

03 Sulfuric acid reacts with sodium hydroxide solution:

$$H_2SO_4 + 2NaOH \rightarrow Na_2SO_4 + 2H_2O$$

In an experiment 25.0 cm³ of 0.100 mol/dm³ sodium hydroxide solution reacts with 27.5 cm³ of sulfuric acid.

03.1 Calculate the concentration of the sulfuric acid in mol/dm³. Give your answer to three significant figures. **[5 marks]**

 **Exam Tip**

Pick out all the key bits of data you need for this and keep it in one place. This will stop you needing to back over the text.

Write down the volume of acid, volume of alkali, concentration of alkali, and ratio of acid to alkali.

03.2 Calculate the concentration of the sulfuric acid in g/dm³. Use your answer to **03.1**. Relative atomic masses A_r: H = 1; S = 32; O = 16 **[2 marks]**

03.3 Calculate the mass of sodium hydroxide in 1 dm³ of 0.100 mol/dm³ solution. Relative atomic masses A_r: Na = 23; O = 16; H = 1 **[2 marks]**

04 Ethanol may be used as a fuel for cars. It can be produced by two different processes. **Table 1** shows these processes.

Table 1

1	$C_2H_4 + H_2O \rightarrow C_2H_5OH$ • occurs at high temperature and high pressure in the presence of phosphoric acid • C_2H_4 (ethene) is obtained from crude oil
2	$C_6H_{12}O_6 \rightarrow 2C_2H_5OH + 2CO_2$ • occurs at 30 °C and atmospheric pressure in the presence of yeast • $C_6H_{12}O_6$ (glucose) is obtained from plants

 **Exam Tip**

The formula for ethanol is given here as C_2H_5OH – this is the same as C_2H_6O.

04.1 Calculate the relative formula mass of ethanol.
The formula of ethanol is C_2H_5OH. **[1 mark]**

04.2 Compare the atom economies of the two processes. **[3 marks]**

04.3 Evaluate the two processes for producing ethanol and suggest which is better for sustainable development. **[6 marks]**

05 Propane gas is used for camping stoves. Its formula is C_3H_8.

05.1 A cylinder contains 6.00 kg of propane. Calculate the volume that this mass of propane would occupy at room temperature and pressure. Give your answer to three significant figures. Relative atomic masses A_r: C = 12; H = 1 **[4 marks]**

05.2 Propane burns in air to make carbon dioxide and water:

$$C_3H_8 + 5O_2 \rightarrow 3CO_2 + 4H_2O$$

Deduce the volume of oxygen that reacts with 50 cm³ of propane. Give your answer in dm³. **[2 marks]**

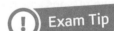 **Exam Tip**

If you do not give your answer to three significant figures you will not get full marks.

05.3 480 g of propane is burnt in air. Calculate the volume of carbon dioxide produced at room temperature and pressure. Give your answer to three significant figures. **[4 marks]**

06 Some students heated 5.6 g of iron with exactly 3.2 g of sulfur. A chemical reaction occurred:

$$Fe(s) + S(s) \rightarrow FeS(s)$$

The mass of product made was 8.2 g.

06.1 What is the maximum theoretical mass of product? Choose **one** answer. **[1 mark]**

2.4 g   5.6 g   8.2 g   8.8 g

Exam Tip

These are one mark questions, so you should be able to do them in under one minute each with little to no calculation.

06.2 What is the actual yield of the reaction? Choose **one** answer.
2.4 g   5.6 g   8.2 g   8.8 g   **[1 mark]**

06.3 What is the atom economy of the reaction? Choose **one** answer.
27 %   68 %   93 %   100 %   **[1 mark]**

07 Titration can be used to deduce the unknown concentration of an acid. Describe the method for carrying out a titration experiment with sodium hydroxide to determine the concentration of hydrochloric acid.   **[6 marks]**

08 Hydrogen may be used as a fuel for cars. It can be made by two processes. **Table 2** shows the two processes.

Table 2

1	$CH_4 + H_2O \rightarrow CO + 3H_2$ • occurs at high temperature • CH_4 (methane) is obtained from sewage or fossil fuels
2	$2H_2O \rightarrow O_2 + 2H_2$ • occurs when electricity is passed through water at room temperature

08.1 Calculate values for the percentage atom economy for process **1**.
Relative atomic masses
A_r: C = 12; O = 16; H = 1   **[3 marks]**

08.2 Suggest how to minimise the environmental impact of process **2**.   **[1 mark]**

08.3 Evaluate the two processes for making hydrogen. Refer to sustainable development in your answer.   **[6 marks]**

09 Some students wanted to find the mass of oxygen and magnesium that react together in this reaction:

$$2Mg(s) + O_2(g) \rightarrow 2MgO(s)$$

Figure 1 shows the apparatus they used.

Figure 1

The students used the following method:

1 Find the mass of some magnesium ribbon and place in the crucible.

2 Heat strongly, removing the lid every now and again.

3 When the magnesium finishes reacting, find the mass of the product.

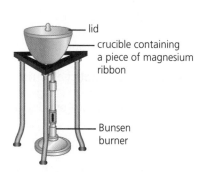

lid
crucible containing a piece of magnesium ribbon
Bunsen burner

09.1 Suggest why the crucible lid is removed every now and again. **[1 mark]**

09.2 **Table 3** shows the students' results.

Table 3

Mass of magnesium ribbon at start in g	1.20
Mass of magnesium oxide at end in g (yield)	1.80

The maximum theoretical mass of magnesium oxide is 2.00 g. Calculate the percentage yield in the students' reaction. **[3 marks]**

09.3 Suggest why the mass of magnesium oxide obtained was less than the maximum theoretical mass. **[1 mark]**

10 A student investigated the volume of nitric acid that reacted with 25 cm³ of a solution of sodium hydroxide.

10.1 The student used 25 cm³ of 0.100 mol/dm³ sodium hydroxide. Their results are shown in **Table 4**.

Identify the outlier. **[1 mark]**

Table 4

Titration number	Volume of nitric acid solution in cm³
1	13.55
2	12.95
3	13.05
4	13.00

10.2 Calculate the mean volume of nitric acid needed to neutralise 25 cm³ of 0.100 mol/dm³ sodium hydroxide. **[1 mark]**

10.3 Write the balanced symbol equation with state symbols for this reaction. **[3 marks]**

10.4 Calculate the concentration of nitric acid in mol/dm³. Give your answer to two significant figures. **[4 marks]**

11 **Figure 2** is an outline of part of the Periodic Table. Each element is represented by a letter. The letters are **not** the chemical symbols of the elements.

Figure 2

11.1 Give the letter of **one** element that is in Group 0. **[1 mark]**

11.2 Give the letters of **two** elements that conduct electricity. **[1 mark]**

11.3 Describe **four** steps in the development of the Periodic Table. **[6 marks]**

Exam Tip

This is a reaction between nitric acid and sodium hydroxide. You'll need to know the formula of both of these.

Then use the general salt equation to work out the products, determine the formula of the salt, and balance the equation. This may seem like a lot but you need to learn and apply them all in an exam.

Exam Tip

You don't need to give the chemical symbols, just the letters shown in **Figure 2**.

12 **Table 5** gives the properties of some elements and compounds. Each substance is represented by a letter.

Table 5

Substance	Melting point in °C	Boiling point in °C	Does it conduct electricity in the solid state?	Does it conduct electricity in the liquid state?
A	−182	−162	no	no
B	801	1465	no	yes
C	650	1110	yes	yes
D	1710	2230	no	no

12.1 Give the letter of the substance that could be methane, CH_4. **[1 mark]**

12.2 Complete the dot and cross diagram for methane shown in **Figure 3**. **[2 marks]**

Figure 3

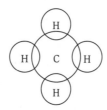

12.3 Explain which substance in **Table 5** could be sodium chloride. **[3 marks]**

12.4 Name and describe the structure and bonding in substance **C**. **[4 marks]**

13 A solution of glucose has a concentration of 55.6 mol/dm³. The formula of glucose is $C_6H_{12}O_6$.

13.1 Calculate the number of moles of glucose in 5.00 dm³ of the solution. **[1 mark]**

13.2 Deduce the concentration of the solution in kg/dm³. Relative atomic masses A_r: C = 12; O = 16; H = 1 **[4 marks]**

13.3 Calculate the mass of glucose in 50.0 cm³ of solution. **[2 marks]**

14 Potassium chloride is a salt. A student has a sample of potassium chloride solution.

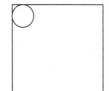

14.1 Describe how the student could separate the potassium chloride from the solution. **[3 marks]**

14.2 Potassium chloride is a solid. Complete the particle diagram to represent solid potassium chloride. Use a circle to represent each potassium chloride particle. **[3 marks]**

14.3 How you can tell that potassium chloride is a compound and not a mixture of potassium and chlorine? **[1 mark]**

 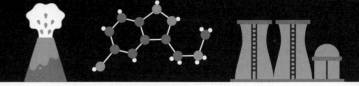

C8 Reactions of metals

Reactions of metals

The **reactivity** of a metal is how chemically reactive it is. When added to water, some metals react very vigorously – these metals have *high* reactivity. Other metals will barely react with water or acid, or won't react at all – these metals have *low* reactivity.

Reactivity series

The reactivity series places metals in order of their reactivity.

Sometimes, for example in the table below, hydrogen and carbon are included in the series, even though they are non-metals.

Reaction with water	Reaction with acid	Reactivity series		Extraction method
		Metal	Reactivity	
fizzes, gives off hydrogen gas	explodes	potassium	**high reactivity**	electrolysis
		sodium		
		lithium		
reacts very slowly	fizzes, gives off hydrogen gas	calcium	Decreasing reactivity	
		magnesium		
		aluminium (carbon)		
		zinc		
		iron		reduction with carbon
no reaction	reacts slowly with warm acid	tin		
		lead (hydrogen)		
		copper		
	no reaction	silver	**low reactivity**	mined from the Earth's crust
		gold		

Metal extraction

Some metals, like gold, are so unreactive that they are found as pure metals in the Earth's crust and can be mined.

Most metals exist as compounds in rock and have to be extracted from the rock. If there is enough metal compound in the rock to be worth extracting it is called an **ore**.

Metals that are less reactive than carbon can be extracted by reduction with carbon. For example:

iron oxide + carbon → iron + carbon dioxide

Metals that are more reactive than carbon can be extracted using a process called **electrolysis**.

Reduction and oxidation

If a substance gains oxygen in a reaction, it has been **oxidised**.

If a substance loses oxygen in a reaction, it has been **reduced**.

For example:

iron + oxygen → iron oxide
iron has been oxidised

iron oxide + carbon → iron + carbon dioxide
iron oxide has been reduced

 **Key terms**

Make sure you can write a definition for these key terms.

displacement   electrolysis   extraction   half equation   ion   ionic equation   metal

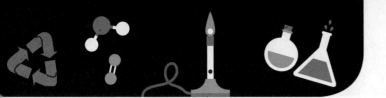

Displacement reactions

In a **displacement** reaction a *more* reactive element takes the place of a *less* reactive element in a compound.

For example:

copper sulfate + iron → iron sulfate + copper

$$CuSO_4(aq) + Fe(s) \rightarrow FeSO_4(aq) + Cu(s)$$

Iron is more reactive than copper, so iron displaces the copper in copper sulfate.

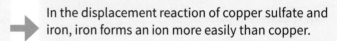

Reactivity and ions

A metal's reactivity depends on how readily it forms an **ion** by losing electrons.

In the displacement reaction of copper sulfate and iron, iron forms an ion more easily than copper.

At the end of the reaction you are left with iron ions, not copper ions.

Ionic equations

When an ionic compound is dissolved in a solution, we can write the compound as its separate ions. For example, $CuSO_4(aq)$ can be written as $Cu^{2+}(aq)$ and $SO_4^{2-}(aq)$.

The displacement reaction of copper sulfate and iron can be written as:

$$Fe(s) + Cu^{2+}(aq) + SO_4^{2-}(aq) \rightarrow Fe^{2+}(aq) + SO_4^{2-}(aq) + Cu(s)$$

The SO_4^{2-} is unchanged in the reaction – it is a **spectator ion**. Spectator ions are removed from the equation to give an **ionic equation**:

$$Fe(s) + Cu^{2+}(aq) \rightarrow Fe^{2+}(aq) + Cu(s)$$

Metals, covalent substances, and solid ionic substances do not split into ions in the ionic equation.

Steps for writing an ionic equation

1 check symbol equation is balanced

2 identify all aqueous ionic compounds

3 write those compounds out as ions

4 remove spectator ions.

Reduction and oxidation: electrons

Oxidation and reduction (**redox** reactions) can be defined in terms of oxygen, but can also be defined as the loss or gain of electrons.

Oxidation is the *loss* of electrons, and reduction is the *gain* of electrons.

In the example displacement reaction:

- iron atoms have been oxidised
- copper ions have been reduced.

Half equations

In the displacement reaction, an iron atom loses two electrons to form a iron ion:

$$Fe(s) \rightarrow Fe^{2+}(aq) + 2e^-$$

A copper ion gains two electrons to form a copper atom:

$$Cu^{2+}(aq) + 2e^- \rightarrow Cu(s)$$

These two equations are called **half equations** – they each show half of the ionic equation.

---

ore   oxidation   reactivity   reactivity series   redox   reduction   spectator ion

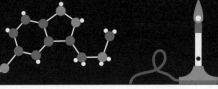

Learn the answers to the questions below then cover the answers column with
a piece of paper and write as many as you can. Check and repeat.

	C8 questions		Answers
1	What does reactivity mean?	Put paper here	how vigorously a substance chemically reacts
2	How can metals be ordered by their reactivity?		by comparing their reactions with water, acid, or oxygen
3	What name is given to a list of metals ordered by their reactivity?		reactivity series
4	In terms of electrons, what makes some metals more reactive than others?	Put paper here	they lose their outer shell electron(s) more easily
5	Why are gold and silver found naturally as elements in the Earth's crust?		they are very unreactive
6	What is an ore?	Put paper here	rock containing enough of a metal compound to be economically worth extracting
7	How are metals less reactive than carbon extracted from their ores?		reduction with carbon
8	In terms of oxygen, what is oxidation?		addition of oxygen
9	In terms of oxygen, what is reduction?	Put paper here	removal of oxygen
10	Why can metals like potassium and aluminium not be extracted by reduction with carbon?		they are more reactive than carbon
11	How are metals more reactive than carbon extracted from their ores?		electrolysis
12	What is a displacement reaction?	Put paper here	a more reactive substance takes the place of a less reactive substance in a compound
13	What is an ionic equation?		equation which gives some substances as ions and has spectator ions removed
14	What type of substance is given as ions in an ionic equation?	Put paper here	ionic compounds in solution (or liquid)
15	What is a spectator ion?		ion that is unchanged in a reaction
16	What is a half equation?		equation that shows whether a substance is losing or gaining electrons
17	In terms of electrons, what is oxidation?	Put paper here	loss of electrons
18	In terms of electrons, what is reduction?		gain of electrons

Now go back and use the questions below to check your knowledge from previous chapters.

Previous questions | Answers

	Previous questions	Answers
1	Why is the actual yield always less than the theoretical yield?	• reaction may be reversible • some of the product can be lost on separation • unexpected side reactions between reactants
2	Where are metals and non-metals located on the Periodic Table?	metals to the left, non-metals to the right
3	What is a unit for concentration?	g/dm^3 or mol/dm^3
4	Why did Mendeleev swap the order of some elements?	to group them by their chemical properties
5	What is a titration?	method used to calculate the concentration of an unknown solution
6	Which formula is used to calculate the mass of a substance from number of moles and M_r?	mass = moles × M_r
7	What is the theoretical yield of a reaction?	maximum mass of the product that could have been produced
8	Why do ionic substances conduct electricity when melted or dissolved?	ions are free to move and carry charge
9	Describe the structure and bonding of small molecules.	small numbers of atoms group together into molecules with strong covalent bonds between the atoms and weak intermolecular forces between the molecules

Put paper here

Maths Skills

Practise your maths skills using the worked example and practice questions below.

Rearranging equations

You need to be able to rearrange and apply many equations in chemistry, for example, the equation to calculate number of moles.

Chemists use moles to describe the relative numbers of particles in a given mass of substance. This can be calculated using:

$$\text{number of moles} = \frac{\text{mass (g)}}{\text{relative atomic mass } (A_r) \textit{ or } \text{relative formula mass } (M_r)}$$

This equation can be rearranged to find the mass of a substance, or the A_r or M_r.

Worked Example

What is the mass of 7.5×10^{-3} moles of aluminium sulfate?

Answer:

Aluminium sulfate = $Al_2(SO_4)_3$

M_r of $Al_2(SO_4)_3$ = (27 × 2) + (32 × 3) + (16 × 12) = 342

Rearrange the equation:

mass = number of moles × M_r

mass = (7.5×10^{-3}) × 342 = 2.565 = 2.6 g

Practice

1 Calculate the relative formula mass of H_2SO_4.

2 Calculate the number of moles of neon atoms in 0.02 g of neon.

3 Calculate the mass of copper sulfate produced through evaporating 1.5 mole copper sulfate solution.

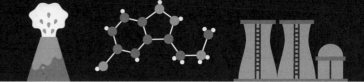
01 This question is about the reactivity series.

01.1 Which of these pairs of substances react together in a displacement reaction?

Tick **one** box. **[1 mark]**

zinc and magnesium chloride solution ☐

zinc and copper chloride solution ☐

iron and zinc sulfate solution ☐

iron and magnesium sulfate solution ☐

> **! Exam Tip**
>
> You are expected to remember the order of elements within the reactivity series. There are lots of rhymes to help you; just pick the one you like the best.

01.2 Name the gas formed when sodium reacts with water. **[1 mark]**

_____

01.3 Lithium also reacts with water. Does lithium react more vigorously or less vigorously with water than sodium?

Draw a circle around **one** answer. **[1 mark]**

more vigorously   less vigorously

> **! Exam Tip**
>
> Identify the position of lithium and sodium on the Periodic Table to help with this question.

01.4 Lithium is added to a solution of copper chloride. Name the substance formed. **[1 mark]**

_____

01.5 Use your answers to questions **01.1** through **01.4** to put copper, lithium, sodium, and zinc in order of reactivity. **[3 marks]**

most reactive   least reactive

_____ > _____ > _____ > _____

02 Iron is found on Earth as iron(III) oxide. To obtain pure iron, iron(III) oxide is reacted with carbon.

02.1 Identify whether the iron is oxidised or reduced. Give a reason for your answer. **[2 marks]**

_____

_____

_____

> **! Exam Tip**
>
> Only write numbers in the boxes, don't try to change the formula of the compounds already given.

02.2 Balance the symbol equation for the extraction of iron from iron(III) oxide. **[1 mark]**

_____ Fe_2O_3 + _____ C → _____ Fe + _____ CO_2

02.3 Explain why you do not need to react gold with carbon to obtain pure gold. **[2 marks]**

_____

_____

_____

03 A student pours some dilute hydrochloric acid into a beaker. The student then adds some pieces of zinc to the acid.

03.1 Describe **one** observation the student would make. **[1 mark]**

03.2 Name the **two** products of the reaction. **[1 mark]**

03.3 Write a balanced chemical equation for the reaction, including state symbols. **[3 marks]**

03.4 Explain whether zinc is oxidised or reduced in the reaction. **[2 marks]**

04 **Figure 1** shows three solutions in test tubes.

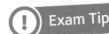

! **Exam Tip**

An observation is what you would _see_ during this reaction.

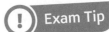

! **Exam Tip**

If you forget the state symbols, you won't get all of the marks.

Figure 1

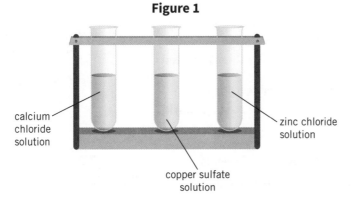

calcium chloride solution

zinc chloride solution

copper sulfate solution

A student added small pieces of metal **X** to each test tube. **Table 1** shows the student's observations.

Table 1

Solution	Observations
calcium chloride	no change
copper sulfate	brown solid forms on the surface of metal **X** and blue colour of solution becomes paler
zinc chloride	grey solid forms on the surface of metal **X**

04.1 Suggest a possible identity of metal **X**. Justify your prediction. **[3 marks]**

! **Exam Tip**

Only giving a metal will not be enough to get all the marks; you must explain why you picked that one.

04.2 Write a balanced chemical equation for the reaction of metal **X** with copper sulfate solution. Use the symbol **X** to represent the metal, and assume it forms **X^{2+}** ions.
Include state symbols in your equation. **[3 marks]**

04.3 Explain how the student could confirm the identify of metal **X**. **[4 marks]**

05 Lead is found naturally as lead sulfide, PbS. Lead sulfide is mixed with other substances in rock. This rock is called lead ore. Lead is extracted from lead ore by the steps below.

    1 Lead sulfide is separated from the substances it is mixed with in lead ore.

    2 Pure lead sulfide is heated with oxygen. This chemical reaction makes lead oxide and sulfur dioxide.

    3 The lead oxide from step **2** is heated with carbon.

05.1 A lead ore contains 25 % lead sulfide.
Calculate the mass of lead in 240 kg of this ore. Give your answer to two significant figures. **[4 marks]**

05.2 Write a balanced symbol equation for step **2**. Include state symbols in your equation. **[3 marks]**

05.3 Explain whether lead oxide is oxidised or reduced in step **3**. **[2 marks]**

05.4 Explain why lead is extracted from its oxide by heating with carbon but aluminium cannot be extracted from its oxide in this way. **[2 marks]**

05.5 Name **one** other metal that can be extracted from its oxide by heating with carbon. **[1 mark]**

06 Compare the displacement reactions of halogens and metals.
In your answer, include ideas about oxidation and reduction and refer to the two equations below.

    Reaction 1 $Cl_2(aq) + 2NaBr(aq) \rightarrow 2NaCl(aq) + Br_2(aq)$

    Reaction 2 $Mg(s) + CuSO_4(aq) \rightarrow MgSO_4(aq) + Cu(s)$ **[6 marks]**

07.1 Which of these metals has the greatest tendency to form positive ions? Choose **one** answer. **[1 mark]**

    iron   lithium   magnesium   zinc

07.2 Name the product formed in the reaction between magnesium and oxygen. **[1 mark]**

07.3 Is magnesium oxidised or reduced in the reaction in **07.2**? Give a reason for your answer. **[2 marks]**

07.4 Magnesium cannot be extracted from the compound formed in **07.2** by reaction with carbon. Explain why. **[2 marks]**

08 A student wants to find the position of nickel in the reactivity series. The student adds small pieces of iron, lead, and nickel to dilute hydrochloric acid and to water. The student's observations are shown in **Table 2**.

Table 2

Metal	Observations on adding the metal to dilute hydrochloric acid	Observations on adding the metal to water and leaving for a few days
iron	bubbles form slowly on the surface of the metal	red-brown flakes form on the surface of the metal
lead	no change	no change
nickel	bubbles form slowly on the surface of the metal	no change

08.1 Use the observations in **Table 2** to deduce the position of nickel in the reactivity series. Justify your decision. **[3 marks]**

This is a common type of question in the exam. Practice it here.

08.2 The student wants to confirm the position of nickel in the reactivity series relative to iron.

Suggest how you could improve the experiment to confirm the position of nickel by using displacement reactions. In your answer, describe and explain the results you would expect. **[4 marks]**

08.3 Nickel will displace copper from a solution of copper(II) sulfate.

Write an ionic equation for the displacement reaction between nickel and copper sulfate. Identify which species is reduced. **[3 marks]**

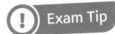

As well as balancing the elements, the charges need to be balanced as well.

09 A student is investigating metal displacement reactions.

The student places small pieces of metal in the depressions of a 3 × 3 white spotting tile. The student then adds small amounts of solutions of metal salts to the metals.

Figure 2 shows the metals and solutions of metal salts that the student used.

Figure 2

magnesium + magnesium chloride solution	zinc + magnesium chloride solution	copper + magnesium chloride solution
magnesium + zinc chloride solution	zinc + zinc chloride solution	copper + zinc chloride solution
magnesium + copper chloride solution	zinc + copper chloride solution	copper + copper chloride solution

09.1 Suggest **one** improvement to the experiment that would prevent the unnecessary use of metals and solutions. **[1 mark]**

09.2 Name the metal that does **not** react with any of the solutions. Explain your choice. [2 marks]

Exam Tip

Find the metals on the reactivity series.

09.3 Write an ionic equation with state symbols for the reaction of magnesium with copper(II) sulfate solution. [3 marks]

09.4 Write the electronic structure of magnesium before and after its reaction with copper(II) sulfate solution. [2 marks]

Exam Tip

This isn't drawing it out, but writing the number of electrons. It starts with 2, 8...

09.5 Suggest **three** advantages of doing this experiment on a spotting tile compared to using test tubes.
Give a reason for each suggestion. [3 marks]

10 The formula of ammonia is NH_3. **Figure 3** is a ball and stick model of an ammonia molecule.

Figure 3

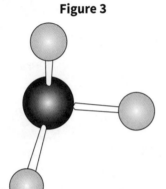

10.1 The ball and stick model is not a true representation of ammonia.
Give **one** reason why. [1 mark]

10.2 Draw a dot and cross diagram to show the covalent bonds in ammonia. [2 marks]

Exam Tip

Covalent bonding has overlapping circles – ionic bonding is the one with square brackets.

10.3 Explain why ammonia does not conduct electricity. [1 mark]

10.4 **Table 3** shows the boiling points and formulas of two compounds.

Table 3

Name of compound	Formula	Boiling point in °C
ammonia	NH_3	−334
hydrazine	N_2H_4	114

Give the state of ammonia at 20 °C. [1 mark]

Exam Tip

You've probably never heard of hydrazine before, but this is just a new context to apply your knowledge.

10.5 Explain why hydrazine has a higher boiling point than ammonia. [2 marks]

11 **Table 4** shows the radii of the atoms in some metals of the reactivity series.

Table 4

Metal	Radius of atoms in metal in nm
potassium	0.231
sodium	0.186
lithium	0.152
calcium	0.197
magnesium	0.160
zinc	0.133
iron	0.126
copper	0.128

A student studies the data in **Table 4** and makes the following conclusion.

'The bigger the atoms of a metal, the higher that metal is in the reactivity series. This is because metals high in the reactivity series lose their electrons more easily.'

11.1 Explain why bigger metal atoms are expected to be more reactive. **[4 marks]**

11.2 Evaluate the extent to which the student's conclusion is true. **[5 marks]**

> ! **Exam Tip**
>
> For an evaluate question you'll need to give reasons for and reasons against your opinion.

12 A student carried out a displacement reaction. They used the following method.

1 Find the mass of the lead oxide powder and carbon powder.

2 Mix the two powders and place in an evaporating basin.

3 Heat strongly.

4 Find the mass of solid remaining after heating and observe the products.

12.1 Lead and its compounds are toxic.

Suggest **two** safety precautions the student should take to reduce the risk of harm from this hazard. **[2 marks]**

12.2 Predict the **two** products of the reaction. **[1 mark]**

12.3 The total mass of solid at the end of the reaction is less than the total mass of reactants at the start. Suggest why. **[1 mark]**

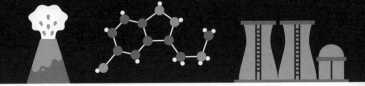

C9 Reactions of acids

Acids and alkalis

Acids are compounds that, when dissolved in water, release H^+ ions. There are three main acids: sulfuric acid H_2SO_4, nitric acid HNO_3, and hydrochloric acid HCl.

Alkalis are compounds that, when dissolved in water, release OH^- ions.

The **pH** scale is a measure of acidity and alkalinity. It runs from 1 to 14.

- Aqueous solutions with pH < 7 are acidic.
- Aqueous solutions with pH > 7 are alkaline.
- Aqueous solutions with pH = 7 are neutral.

Logarithmic scales

The pH scale tells you how many H^+ ions there are in the solution. The *more* H^+ ions are present, the *lower* the pH. It is a logarithmic scale, which means that an increase of 1 on the pH scale is equal to a decrease of 10× the number of H^+ ions in solution.

For example, an acid with a pH of 3 has:

- 100 × *fewer* H^+ ions in solution than an acid with pH 1
- 10 × *fewer* H^+ ions than pH 2
- 10 × *more* H^+ ions than pH 4
- 100 × *more* H^+ ions than pH 5

Indicators

Indicators can show if something is an acid or an alkali.

- **Universal indicator** can also tell us the approximate pH of a solution.
- Electronic pH probes can give us the exact pH of a solution.

Strong and weak acids

Sulfuric acid, nitric acid, and hydrochloric acid, are all **strong acids**. This means that, when dissolved in water, every molecule splits up into ions – they are completely ionised:

- $H_2SO_4(aq) \rightarrow 2H^+(aq) + SO_4^{2-}(aq)$
- $HNO_3(aq) \rightarrow H^+(aq) + NO_3^-(aq)$
- $HCl(aq) \rightarrow H^+(aq) + Cl^-(aq)$

Ethanoic acid, citric acid, and carbonic acid are **weak acids**. This means that only a percentage of their molecules split up into ions when dissolved in water – they are partially ionised.

For a given concentration, the *stronger* the acid, the *lower* the pH.

Concentrated and dilute acids

Concentration tells us how much of a substance there is dissolved in water:

- more concentrated acids have lots of acid in a small volume of water
- less concentrated acids (dilute acids) have little acid in a large volume of water.

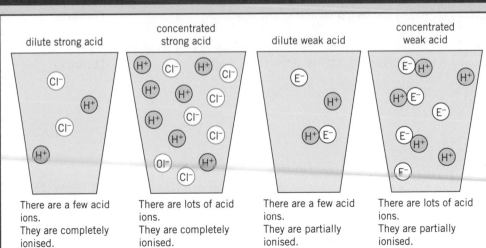

Salts

When acids react with metals or metal compounds, they form salts. A salt is a compound where the hydrogen from an acid has been replaced by a metal. For example nitric acid, HNO_3, reacts with sodium to form $NaNO_3$. The H in nitric acid is replaced with Na.

The table shows how to name salts.

Acid	hydrochloric acid	sulfuric acid	nitric acid
Formula	HCl	H_2SO_4	HNO_3
Ions formed in solution	H^+ and Cl^-	$2H^+$ and SO_4^{2-}	H^+ and NO_3^-
Type of salt formed	metal chloride	metal sulfate	metal nitrate
Sodium salt example	sodium chloride, NaCl	sodium sulfate, Na_2SO_4	sodium nitrate, $NaNO_3$

Reactions of acids

neutralisation reactions

Reactions of acids with metals

Acids react with some metals to form salts and hydrogen gas.

magnesium + hydrochloric acid → sodium chloride + hydrogen

Reactions of acids with metal hydroxides

Acids react with metal hydroxides to form salts and water.

hydrochloric acid + sodium hydroxide → sodium chloride + water

The ionic equation for this reaction is always:

$$H^+(aq) + OH^-(aq) \rightarrow H_2O(l)$$

Reactions of acids with metal oxides

Acids react with metal oxides to form salts and water.

hydrochloric acid + sodium oxide → sodium chloride + water

Reactions of acids with metal carbonates

Acids react with metal carbonates to form a salt, water, and carbon dioxide.

hydrochloric acid + sodium carbonate → sodium chloride + water + carbon dioxide

Redox

The reaction of acids with metals is a redox reaction:
- the metal loses electrons – it is oxidised
- hydrogen gains an electron – it is reduced.

Alkalis and bases

Bases neutralise acids to form water in **neutralisation** reactions. Some metal hydroxides dissolve in water to form alkaline solutions, called alkalis.

Some metal oxides and metal hydroxide do not dissolve in water. They are **bases**, but are not alkalis.

Crystallisation

You can produce a solid salt from an insoluble base by **crystallisation**.

The experimental method is:
1 Choose the correct acid and base to produce the salt.
2 Put some of the dilute acid into a flask. Heat gently with a Bunsen burner.
3 Add a small amount of the base and stir.
4 Keep adding the base until no more reacts – the base is now in excess.
5 Filter to remove the unreacted base.
6 Add the remaining solution to an evaporating dish.
7 Use a water bath or electric heater to evaporate the water. The salt crystals will be left behind.

Key terms

Make sure you can write a definition for these key terms.

acid   alkali
base   concentrated
crystallisation   dilute
indicator   ionise
logarithmic   neutral
neutralisation   pH
salt   strong
universal indicator
weak

Learn the answers to the questions below then cover the answers column with a piece of paper and write as many as you can. Check and repeat.

	C9 questions		Answers
1	In terms of pH, what is an acid?	Put paper here	a solution with a pH of less than 7
2	In terms of pH, what is a neutral solution?		a solution with a pH of 7
3	In terms of H^+ ions, what is an acid?		a substance that releases H^+ ions when dissolved in water
4	How is the amount of H^+ ions in a solution related to its pH?	Put paper here	the more H^+ ions, the lower the pH
5	What are the names and formulae of three main acids?		hydrochloric acid, HCl; sulfuric acid, H_2SO_4; nitric acid, HNO_3
6	How do you measure the pH of a substance?		universal indicator or pH probe
7	What is a strong acid?	Put paper here	an acid where the molecules or ions completely ionise in water
8	What is a weak acid?		an acid where the molecules or ions partially ionise in water
9	What is a salt?	Put paper here	compound formed when a metal ion takes the place of a hydrogen ion in an acid
10	Which type of salts do sulfuric acid, hydrochloric acid, and nitric acid form?		sulfates, chlorides, nitrates
11	What are the products of a reaction between a metal and an acid?		salt + hydrogen
12	What are the products of a reaction between a metal hydroxide and an acid?	Put paper here	salt + water
13	What are the products of a reaction between a metal oxide and an acid?		salt + water
14	What are the products of a reaction between a metal carbonate and an acid?	Put paper here	salt + water + carbon dioxide
15	What is a base?		substance that reacts with acids in neutralisation reactions
16	What is an alkali?		substance that dissolves in water to form a solution above pH 7
17	What is a neutralisation reaction?	Put paper here	a reaction between an acid and a base to produce water
18	What is the ionic equation for a reaction between an acid and an alkali?		$H^+(aq) + OH^-(aq) \rightarrow H_2O(l)$
19	How can you obtain a solid salt from a solution?	Put paper here	crystallisation
20	When an acid reacts with a metal, which species is oxidised?		the metal
21	When an acid reacts with a metal, which species is reduced?		hydrogen

Now go back and use the questions below to check your knowledge from previous chapters.

Previous questions | Answers

#	Previous questions		Answers
1	What volume does one mole of any gas occupy at room temperature and pressure?	Put paper here	24 dm³ or 24 000 cm³
2	Which formula is used to calculate the number of moles from mass and Mr?		$moles = \dfrac{mass}{M_r}$
3	Which formula is used to calculate concentration from mass and volume?	Put paper here	$concentration\ (g/dm^3) = \dfrac{mass\ (g)}{volume\ (dm^3)}$
4	What is an ore?		rock containing enough of a metal compound to be economically worth extracting
5	What is a displacement reaction?		a more reactive substance takes the place of a less reactive substance in a compound
6	Why does the reactivity of the alkali metals increase down the group?	Put paper here	they are larger atoms, so the outermost electron is further from the nucleus, meaning there are weaker electrostatic forces of attraction and more shielding between the nucleus and outer electron, and it is easier to lose the electron
7	What is a half equation?		equation that shows whether a substance is losing or gaining electrons
8	How is percentage atom economy calculated?	Put paper here	$\dfrac{M_r\ of\ useful\ product}{M_r\ of\ all\ products} \times 100$
9	How do the melting points of the noble gases change down the group?		increase (higher melting point)

Required Practical Skills

Practise answering questions on the required practicals using the example below.
You need to be able to apply your skills and knowledge to other practicals too.

Making salts	Worked Example	Practice
This required practical tests whether you can safely separate and purify a chemical mixture, to produce a salt. You will need to be able to describe the uses of filtration, evaporation, and crystallisation to make pure, dry samples of soluble salts. Exam questions can ask about the production of any salt, not just the examples you are familiar with. You should also be able to describe how substances can be tested for purity.	A salt is formed when an acid reacts with a base. 1 Write a word equation for the production of magnesium nitrate. magnesium carbonate + nitric acid → magnesium nitrate + water + carbon dioxide 2 Identify one hazard of working with sulfuric acid, and describe two ways to prevent this hazard. Concentrated sulfuric acid is corrosive, so can cause burns. To prevent any hazards, eye protection should be worn at all times and dilute sulfuric acid should be used.	1 Describe the colour changes that occur when copper oxide is mixed with sulfuric acid. 2 A student made a sample of copper sulfate by reacting copper oxide with sulfuric acid. After evaporating the copper sulfate solution and leaving it to crystallise, the student found two different crystals in the basin. Small white crystals formed around the edges and larger blue crystals formed in the middle. Name the two types of crystal. 3 The student used filter paper in the production of their copper sulfate sample. Describe the function of the filter paper.

Practice

Exam-style questions

01 This question is about the pH scale and neutralisation.

01.1 Draw **one** line from each pH to a solution that might have this pH.

[3 marks]

pH	Solution that might have this pH
	weak acid
1	strong acid
5	neutral solution
9	weak alkali
	strong alkali

Exam Tip

Not all of the boxes on the right hand side will have lines going into them.

01.2 Which solution has the lowest pH?

Tick **one** box.   [1 mark]

1 mol/dm³ solution of ammonium hydroxide ☐

1 mol/dm³ solution of citric acid ☐

1 mol/dm³ solution of ethanoic acid ☐

1 mol/dm³ solution of hydrochloric acid ☐

Exam Tip

The strongest acid will have the lowest pH.

Exam Tip

It is important that you can recall the list of strong and weak acids from the specification.

01.3 Give the name and formula of the ion that is produced by acids in aqueous solution.   [2 marks]

Name: _____

Formula: _____

01.4 Give the name of the product that is always formed when an acid reacts with a base.   [1 mark]

_____

02 A student has two solutions of hydrochloric acid.
Table 1 shows the pH of each solution.

Table 1

Solution	pH
A	1
B	2

02.1 Which statement about the two solutions is correct?
Tick **one** box. **[1 mark]**

The H^+ concentration in solution **B** is ten times the H^+ concentration in solution **A**. ☐

The H^+ concentration in solution **A** is ten times the H^+ concentration in solution **B**. ☐

The H^+ concentration in solution **A** is half the H^+ concentration in solution **B**. ☐

The H^+ concentration in solution **A** is double the H^+ concentration in solution **B**. ☐

> **! Exam Tip**
> Remember pH is a log scale, so it goes up in jumps of 10.

02.2 What are the products of the reaction of copper carbonate with hydrochloric acid? **[1 mark]**

_____

02.3 Give the chemical formula of the salt produced in the reaction between hydrochloric acid and copper carbonate. **[1 mark]**

_____

> **! Exam Tip**
> There will be **three** products from this reaction – it is important that you can recall and apply all of the general salt equations in chemistry.

03 A student carried out a titration between hydrochloric acid and sodium hydroxide. This is the method used:
1 Fill a burette with acid.
2 Transfer 25.0 cm³ of sodium hydroxide to a conical flask.
3 Add a few drops of indicator to the sodium hydroxide.
4 Add acid from the burette to the flask until the indicator changes colour.
5 Repeat the procedure.

03.1 Name the piece of apparatus used to transfer the sodium hydroxide in step **2**. **[1 mark]**

03.2 Suggest **two** improvements to step **4**. **[2 marks]**

03.3 Describe how the student will know when to stop repeating the procedure. **[1 mark]**

> **! Exam Tip**
> This has to be the best piece of scientific equipment for the job, not anything that can hold liquid.

For answers and more practice questions visit www.oxfordrevise.com/scienceanswers | Even more practice and interactive revision quizzes are available on kerboodle | C9 Practice 93

03.4 **Table 4** shows the results the student obtained.

Table 4

	Titration 1	Titration 2	Titration 3	Titration 4	Titration 5
Initial burette reading in cm³	1.20	24.50	0.80	21.10	20.50
Final burette reading in cm³	24.50	45.70	21.10	42.35	40.75
Volume of acid added in cm³	23.30	21.20	21.30	21.25	20.25

Use the student's results to calculate the mean volume of acid added. Give your answer to **one** decimal place. **[1 mark]**

03.5 The equation for the reaction is:

$$NaOH(aq) + HCl(aq) \rightarrow NaCl(aq) + H_2O(l)$$

The concentration of acid used was 0.200 mol/dm³. Calculate the concentration of sodium hydroxide in mol/dm³. Give your answer to three significant figures. **[4 marks]**

Exam Tip

Find and pull out all the key bits of data you need for this and write it in one place. This will stop you needing to read back over the text.

volume of acid = mean titre

volume of alkali = 25.0 cm³

concentration of acid = 0.200 mol/dm³

ratio of acid to alkali = 1:1

04 Some students wanted to make zinc chloride, $ZnCl_2$. This is the method they used:

1. Add excess zinc oxide to hydrochloric acid.

2. Filter the mixture.

3. Pour the filtrate into an evaporating basin.

4. Heat the evaporating basin and its contents until all the water has evaporated.

04.1 Suggest **two** improvements to step **1** to speed up the reaction. **[2 marks]**

Exam Tip

This is similar to a required practical that you should have done in class – the change in reactants might worry you at first, but the method is exactly the same.

04.2 Describe how the students would know when to stop adding zinc oxide in step **1**. **[1 mark]**

04.3 Give a reason why the mixture is filtered in step **2**. **[1 mark]**

04.4 Write down what the students should use to heat the evaporating basin and its contents in step **4**. **[1 mark]**

Exam Tip

In exams you will frequently use other parts of the question to find the information you need to answer later parts. If you get used to this now it will be easier in the exam.

04.5 Suggest an improvement to step **4** to make larger zinc chloride crystals. **[2 marks]**

04.6 Write a balanced chemical equation for the reaction that occurs in step **1**. **[3 marks]**

04.7 In step **1**, the students had 25.0 cm³ of hydrochloric acid. The concentration of hydrochloric acid was 0.50 mol/dm³.

Calculate the maximum mass of zinc chloride that could be made. Give your answer to **one** significant figure. **[6 marks]**

Exam Tip

This is the reaction between zinc oxide and hydrochloric acid.

05 Magnesium reacts with dilute sulfuric acid.

$$Mg(s) + H_2SO_4(aq) \rightarrow MgSO_4(aq) + H_2(g)$$

05.1 Explain which substance is oxidised in the reaction of magnesium with sulfuric acid. Include a full balanced ionic equation, with state symbols, in your answer. **[5 marks]**

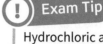

(!) Exam Tip

Hydrochloric acid is HCl, so one product will be the same and one different.

05.2 Magnesium also reacts with hydrochloric acid. Give the formulae of the **two** products of the reaction. **[2 marks]**

06 This question is about strong and weak acids.

06.1 Name **one** weak acid and **one** strong acid. **[2 marks]**

06.2 A solution of hydrochloric acid with a hydrogen ion concentration of 1×10^{-3} mol/dm³ has a pH of 3. Deduce the pH of hydrochloric acid with a hydrogen ion concentration of 1×10^{-5} mol/dm³. Explain your answer. **[3 marks]**

(!) Exam Tip

This may seem like a really hard question but remember that pH is a log scale. 1 pH unit = a hydrogen ion concentration change by a factor of 10.

06.3 A technician made up solutions of two acids, represented by the formulae HX and HZ.

- Solution **A** contains a weak acid of formula HX.

  20 % of the HX molecules are dissociated in solution.

  The concentration of the solution is 5 mol/dm³.

- Solution **B** contains a strong acid of formula HZ.

  100 % of the HZ is dissociated in solution.

  The concentration of the solution is 2 mol/dm³.

Deduce which solution has the lower pH, solution **A** or solution **B**. Explain your answer. **[4 marks]**

07 Sulfuric acid reacts with magnesium to form magnesium sulfate.

07.1 The chemical formula of sulfuric acid is H_2SO_4. What is the charge of the sulfate ion, SO_4, in sulfuric acid? **[1 mark]**

07.2 Give the chemical formula of magnesium sulfate. **[1 mark]**

(!) Exam Tip

Sulfuric acid is a neutral compound so the charges on the ions within sulfuric acid must be equal to zero.

07.3 Manganese is another metal that forms 2+ ions. Manganese reacts with hydrochloric acid. Give the formula of manganese chloride. **[1 mark]**

07.4 Name the gas released when magnesium and manganese react with acids. **[1 mark]**

08 A student wanted to make large copper sulfate crystals from copper hydroxide and an acid.

08.1 Name the acid the student should use. **[1 mark]**

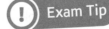

(!) Exam Tip

The second part of the salt name should point you towards the acid used.

08.2 Describe how the student could make a sample of copper sulfate crystals from copper hydroxide and the acid. In your answer:

- Name the pieces of apparatus required.
- Give a reason for each step. **[6 marks]**

08.3 Write a balanced symbol equation with state symbols for the reaction between the acid and copper hydroxide, $Cu(OH)_2$. **[3 marks]**

08.4 The student used $30\,cm^3$ of $32.5\,g/dm^3$ copper hydroxide solution. Calculate the moles of copper hydroxide that were used in the reaction. $M_r(Cu(OH)_2) = 97.5$ **[4 marks]**

08.5 Calculate the maximum mass of copper sulfate that the student could produce. Give your answer to two significant figures. $M_r(CuSO_4) = 159.5$ **[5 marks]**

09 A student made some zinc nitrate crystals by reacting zinc carbonate with nitric acid. The equation for the reaction is:

$$ZnCO_3(s) + 2HNO_3(aq) \rightarrow Zn(NO_3)_2(aq) + CO_2(g) + H_2O(l)$$

The mass of zinc nitrate made was $9.45\,g$.

09.1 Calculate the mass of zinc carbonate that reacted to make $9.45\,g$ of zinc nitrate. **[5 marks]**

09.2 To make the zinc nitrate, the student started with some acid in a conical flask. The student then added zinc carbonate, a little at a time, until some remained unreacted. Explain how the pH of the mixture in the conical flask changed as the student added zinc carbonate. Include a possible pH value for the acid at the start of the reaction. **[3 marks]**

10 This question is about the pH scale. A student measured the pH of some solutions. **Table 3** shows the results the student obtained.

Table 3

Solution	pH
A	7
B	2
C	10
D	5
E	12

10.1 Name **two** ways of measuring the pH of a solution. **[2 marks]**

10.2 Give the letter of the neutral solution. **[1 mark]**

10.3 Give the letter of the most alkaline solution. **[1 mark]**

10.4 Give the letter of the solution that has the highest concentration of hydrogen ions, H^+. **[1 mark]**

10.5 Some alkali is added to solution **A**. Write down whether the pH increases, decreases, or stays the same. **[1 mark]**

11 A student has a solution of $20\,g/dm^3$ sodium hydroxide and sulfuric acid of an unknown concentration.

11.1 Write a balanced symbol equation for the reaction between sodium hydroxide and sulfuric acid. Include state symbols. **[3 marks]**

11.2 Describe a titration method by which the student can determine the concentration of sulfuric acid. **[6 marks]**

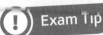

11.3 The student determines that the concentration of sulfuric acid was 29.4 g/dm³. The student used 25 cm³ of sulfuric acid in their titration. Calculate the number of moles of sodium hydroxide that was needed to neutralise 25 cm³ of sulfuric acid. **[6 marks]**

11.4 Calculate the volume of sodium hydroxide that was required to neutralise 25 cm³ of sulfuric acid. **[6 marks]**

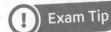

12 **Table 4** shows the pH of some acid solutions. Two of the solutions contain citric acid and two of the solutions contain hydrochloric acid.

Table 4

Solution	Concentration in mol/dm³	pH
W	0.10	5
X	0.10	3
Y	1.0	3
Z	1.0	1

12.1 Give the letter of the solution that has the highest hydrogen ion concentration. **[1 mark]**

12.2 Determine the letters of the **two** solutions of citric acid. Justify your decision. **[4 marks]**

12.3 Hydrochloric acid is a strong acid. Citric acid is a weak acid. Describe the difference between a weak acid and a strong acid. **[1 mark]**

12.4 Name **one** other weak acid and **one** other strong acid. **[2 marks]**

12.5 A student states that:

'The pH of a weak acid is always higher than the pH of a strong acid.'

Evaluate this statement. **[6 marks]**

13 This question is about the elements in Group 0 of the Periodic Table.

13.1 Give the general name for the elements in Group 0. **[1 mark]**

13.2 An argon atom has 18 electrons.

Write the electronic structure for argon. **[1 mark]**

13.3 Explain why the Group 0 elements are unreactive. **[1 mark]**

13.4 Describe the trend in boiling points from top to bottom of Group 0. **[1 mark]**

14 An ion has the formula $_{31}^{69}\text{Ga}^{3+}$.

14.1 Give the number of protons in the ion. **[1 mark]**

14.2 Give the number of neutrons in the ion. **[1 mark]**

14.3 Give the number of electrons in the ion. **[1 mark]**

14.4 Give the name of **one** other element that is in the same group of the Periodic Table as gallium. **[1 mark]**

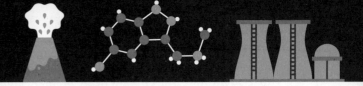

C10 Electrolysis

Electrolysis

In the process of **electrolysis**, an electric current is passed through an **electrolyte**. An electrolyte is a liquid or solution that contains ions and so can conduct electricity. This causes the ions to move to the **electrodes**, where they form pure elements.

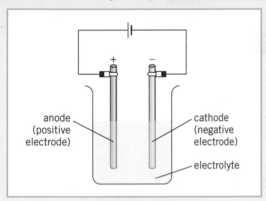

anode (positive electrode)

cathode (negative electrode)

electrolyte

Electrolysis of molten compounds

Solid ionic compounds do not conduct electricity as the ions cannot move. To undergo electrolysis they must be molten or dissolved, so the ions are free to move.

When an ionic compound is molten:

- The positive metal ions are *attracted* to the **cathode**, where they will *gain* electrons to form the pure metal
- The negative non-metal ions are *attracted* to the **anode**, where they will *lose* electrons and become the pure non-metal.

For example, molten sodium chloride, NaCl, can undergo electrolysis to form sodium at the cathode and chlorine at the anode.

Half equations

sodium chloride → sodium + chlorine

$$2NaCl(l) \rightarrow 2Na(s) + Cl_2(g)$$

- at the cathode:

$$2Na^+(l) + 2e^- \rightarrow 2Na(s)$$

- at the anode:

$$2Cl^-(l) \rightarrow Cl_2(g) + 2e^-$$

Electrolysis of aqueous solutions

Solid ionic compounds can also undergo electrolysis when dissolved in water.

- It requires less energy to dissolve ionic compounds in water than it does to melt them.
- However, in the electrolysis of solutions, the pure elements are not always produced. This is because the water can also undergo ionisation:

$$H_2O(l) \rightarrow H^+(aq) + OH^-(aq)$$

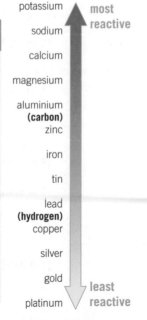

most reactive

potassium
sodium
calcium
magnesium
aluminium
(carbon)
zinc
iron
tin
lead
(hydrogen)
copper
silver
gold
platinum

least reactive

Products at the anode

In In the electrolysis of a solution, if the non-metal contains oxygen then oxygen gas is formed at the anode:

- The $OH^-(aq)$ ions formed from the ionisation of water are attracted to the anode.
- The $OH^-(aq)$ ions lose electrons to the anode and form oxygen gas.
- $4OH^-(aq) \rightarrow O_2(g) + 2H_2O(l) + 4e^-$

If the non-metal ion is a halogen, then the halogen gas is formed at the anode.

- $2Cl^-(aq) \rightarrow Cl_2(g) + 2e^-$

Products at the cathode

In the electrolysis of a solution, if the metal is *more* **reactive** than hydrogen then hydrogen gas is formed at the cathode:

- The $H^+(aq)$ ions from the ionisation of water are attracted to the cathode and react with it.
- The $H^+(aq)$ ions gain electrons from the cathode and form hydrogen gas.
- $2H^+(aq) + 2e^- \rightarrow H_2(g)$
- The metal ions remain in solution.

Electrolysis of aluminium oxide

Electrolysis can be used to extract metals from their ionic compounds.

Electrolysis is used if the metal is more reactive than carbon.

Aluminium is extracted from aluminium oxide by electrolysis.

1 The aluminium oxide is mixed with a substance called **cryolite**, which lowers the melting point.
2 The mixture is then heated until it is molten.
3 The resulting molten mixture undergoes electrolysis.

aluminium oxide   →   aluminium   +   oxygen

$$2Al_2O_3(l) \quad \rightarrow \quad 4Al(l) \quad + \quad 3O_2(g)$$

cathode: pure aluminium is formed
$$Al^{3+}(l) + 3e^- \rightarrow Al(l)$$

anode: oxygen is formed
$$2O^{2-}(l) \rightarrow O_2(g) + 4e^-$$

In the electrolysis of aluminium, the anode is made of graphite.

The graphite reacts with the oxygen to form carbon dioxide and so slowly wears away. It therefore needs to be replaced frequently.

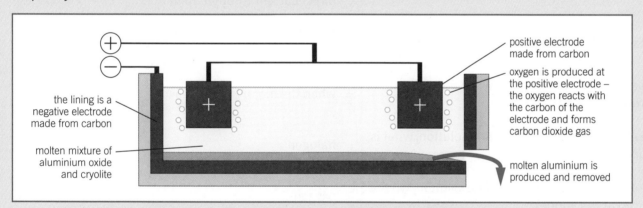

positive electrode made from carbon

oxygen is produced at the positive electrode – the oxygen reacts with the carbon of the electrode and forms carbon dioxide gas

the lining is a negative electrode made from carbon

molten mixture of aluminium oxide and cryolite

molten aluminium is produced and removed

 **Revision tips**

Extraction of aluminium by electrolysis makes a good six mark question. It's a bit different from the other example of electrolysis that you need to learn. Make sure you can do the half equations, know the key words, and know what the electrodes are made of.

 **Revision tips**

In an exam, don't PANIC.

Here is an easy way to remember which electrode is which:

Positive
Anode
Negative
Is
Cathode

 **Key terms**

Make sure you can write a definition for these key terms.

anode   cathode   cryolite   electrode

electrolysis   electrolyte   reactivity

 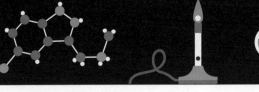

Learn the answers to the questions below then cover the answers column with a piece of paper and write as many as you can. Check and repeat.

C10 questions | Answers

#	Question	Answer
1	What is electrolysis?	process of using electricity to extract elements from a compound
2	What is the name of the positive electrode?	anode
3	What is the name of the negative electrode?	cathode
4	What is an electrolyte?	liquid or solution that contains ions and so can conduct electricity
5	Where are metals formed?	cathode
6	Where are non-metals formed?	anode
7	How can ionic substances be electrolysed?	by melting or dissolving them, and then passing a direct current through them
8	Why can solid ionic substances not be electrolysed?	they do not conduct electricity, or the ions cannot move
9	In the electrolysis of solutions, when is the metal *not* produced at the cathode?	when the metal is more reactive than hydrogen
10	In the electrolysis of a metal halide solution, what is produced at the anode?	halogen
11	In the electrolysis of a metal sulfate solution, what is produced at the anode?	oxygen
12	What is the half equation for the ionisation of water?	$H_2O(l) \rightarrow H^+(aq) + OH^-(aq)$
13	What metals are extracted from ionic compounds by using electrolysis?	metals that are more reactive than carbon
14	In the electrolysis of aluminium oxide, why is the aluminium oxide mixed with cryolite?	to lower the melting point
15	In the electrolysis of aluminium oxide, what are the anodes made of?	graphite
16	In the electrolysis of aluminium oxide, why do the anodes need to be replaced?	they react with the oxygen being formed

Put paper here

Now go back and use the questions below to check your knowledge from previous chapters.

Previous questions

Answers

	Previous questions	Answers
1	What are the names and formulae of three main acids?	hydrochloric acid, HCl; sulfuric acid, H_2SO_4; nitric acid, HNO^3
2	How are metals more reactive than carbon extracted from their ores?	electrolysis
3	How big are fine particles?	100–2500 nm
4	What is a mole?	mass of a substance that contains 6.02×10^{23} particles
5	What is relative formula mass?	the sum of the relative atomic masses of each atom in a substance
6	How big are coarse particles?	2.5×10^{-6} to 1×10^{-5} m
7	What is the ionic equation for a reaction between an acid and an alkali?	$H^+(aq) + OH^-(aq) \rightarrow H_2O(l)$
8	What is a limiting reactant?	the reactant that is completely used up in a chemical reaction
9	Why are gold and silver found naturally as elements in the Earth's crust?	they are very unreactive

Put paper here

Put paper here

Put paper here

Required Practical Skills

Practise answering questions on the required practicals using the example below. You need to be able to apply your skills and knowledge to other practicals too.

Electrolysis	Worked Example	Practice
You need to be able to describe the method of electrolysis, and label the experimental set-up for electrolysis. Electrolysis uses electricity to break ionic compounds down into simpler compounds or elements. Metals or hydrogen are made at the negative electrode, and non-metal molecules are made at the positive electrode. You will need to be able to apply the principles of electrolysis to any example, as many solutions can undergo electrolysis. This includes predicting the products of electrolysis for different solutions, identifying which ions move to each electrode, and writing equations for the reactions at the two electrodes.	The electrolysis of aqueous sodium chloride gives three products. Identify these products and state how we can test for them. **Answer**: The three products are chlorine gas (Cl_2) hydrogen gas (H_2) and sodium hydroxide solution (NaOH). To test for hydrogen gas, collect the gas in a test tube and insert a glowing splint – it should burn with a squeaky pop noise. To test for chlorine gas, collect the gas in a test tube and insert damp litmus paper – the litmus paper will bleach white. Sodium hydroxide can be tested for using universal indicator – the solution will turn purple as sodium hydroxide is an alkali.	**1** State what you would observe at each electrode during the electrolysis of copper(II) chloride. **2** Give the products of the electrolysis of sodium sulfate. **3** Explain why the electrodes must not touch each other during electrolysis.

Practice

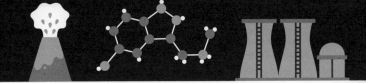

01 A student investigated the electrolysis of sodium chloride solution.
Figure 1 shows the apparatus used.

Figure 1

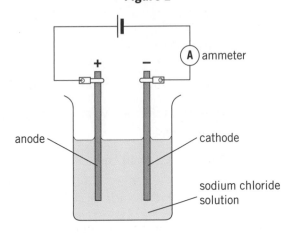

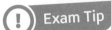

> **! Exam Tip**
>
> It's not sodium. Use the reactivity series and the formula of salty water to work out the other product.

01.1 Name the substance that is produced at the cathode. **[1 mark]**

_____

01.2 Write a half equation, including state symbols, for the reaction that occurs at the anode. **[3 marks]**

_____

01.3 The student wanted to investigate if changing the concentration of sodium chloride solution affects the current that flows.

Table 1 shows the student's results.

Table 1

Concentration of sodium chloride solution in mol/dm³	Current in amps
0.2	0.20
0.4	0.33
0.6	0.43
0.8	0.47
1.0	0.52

Identify the independent variable and the dependent variable in the investigation. **[2 marks]**

independent variable: _____

dependent variable: _____

01.4 Suggest **one** control variable in the investigation. **[1 mark]**

_____

> **! Exam Tip**
>
> 'Independent' is the one we change and 'dependent' is the one we measure. A good way to remember this is that your results _depend_ on the dependent variable.

01.5 Plot the data from **Table 1** on **Figure 2**.

Draw a line of best fit. **[3 marks]**

Figure 2

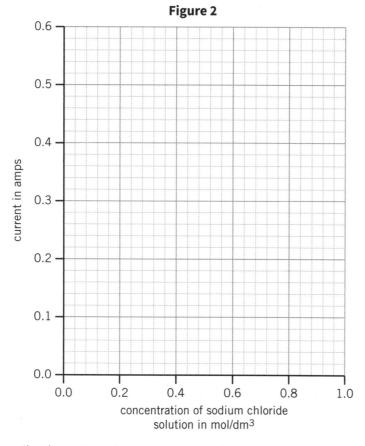

> **(!) Exam Tip**
>
> Use crosses to plot your points because this clearly shows the examiners which point you are aiming for. Circles can easily be misinterpreted as they can cover a range of points or be too small to be seen by the examiner. Crosses are the best way to ensure you get the mark.

01.6 Describe the pattern shown on your graph.

Suggest **one** reason for this pattern. **[2 marks]**

Pattern: _____

_____

Reason: _____

_____

> **(!) Exam Tip**
>
> Use data from the graph to support your reason.

02 Molten zinc chloride is electrolysed using inert electrodes.

02.1 Name the electrode that positively charged ions move towards during electrolysis. **[1 mark]**

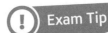

_____

02.2 What are the products at the anode and cathode? **[1 mark]**

Tick **one** box.

anode	cathode	
zinc	chlorine	☐
chlorine	zinc	☐
zinc	hydrogen	☐
chlorine	hydrogen	☐

Exam Tip

The first step is to work out the charges on the ions within zinc chloride.

02.3 Explain why solid zinc chloride cannot be used for electrolysis. **[3 marks]**

_____

_____

_____

_____

02.4 The symbol equation for the reaction is:

$$ZnCl_2 \rightarrow Zn + Cl_2$$

Complete the symbol equation by adding state symbols. **[1 mark]**

03 Potassium is extracted from its ores by electrolysis.

03.1 Suggest why electrolysis is used to extract potassium. **[1 mark]**

03.2 In the electrolysis of molten potassium sulfate, name the electrode that solid potassium metal will form at. **[1 mark]**

03.3 The electrolysis of molten potassium sulfate is an expensive industrial process. Give a reason why. **[1 mark]**

03.4 Aqueous potassium sulfate solution can also be electrolysed. Write the half equations for the electrolysis of aqueous potassium sulfate. **[6 marks]**

Exam Tip

Look at the reactivity series to help you answer this.

03.5 Suggest why an aqueous solution of potassium sulfate cannot be used to extract potassium. **[1 mark]**

04 **Figure 3** shows an electrolysis cell for the industrial extraction of aluminium.

Figure 3

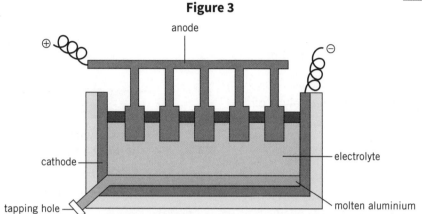

04.1 Explain why aluminium cannot be extracted by heating its ore with carbon. **[1 mark]**

04.2 Name the material that the anode and cathode are made from. **[1 mark]**

04.3 Explain why the anode must be replaced regularly. **[1 mark]**

04.4 Name the **two** substances that are mixed together in the electrolyte. **[2 marks]**

04.5 Write a half equation, including state symbols, for the reaction that occurs at the cathode. **[3 marks]**

04.6 Suggest why industrial aluminium electrolysis cells are often sited near power stations that generate electricity from renewable sources. **[1 mark]**

05.1 Define the term inert electrode. **[2 marks]**

05.2 The charge on a lead ion is 2+. Deduce the formula of lead bromide. **[1 mark]**

05.3 An electrolysis reaction happens when electricity is passed through molten lead bromide using inert electrodes. Describe what happens in this reaction. Include in your answer the name of the products of the electrolysis reaction and an explanation of how the products are made **[6 marks]**

06 A student investigates electrolysis cells. **Figure 4** shows the apparatus used.

! Exam Tip

Think about the element that the electrodes are made from.

! Exam Tip

Make sure the number of electrons matches the charge on the ions, and that the equation is balanced.

! Exam Tip

Split your answer into two parts: what happens at the anode and then what happens at the cathode.

Figure 4

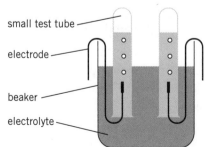

The student used the following method:

1 Set up the electrolysis cell as shown, with a battery to supply the current.

2 Pour the first electrolyte into the beaker.

3 Switch on the current and record any observations.

06.1 Suggest a reason for investigating the electrolysis of aqueous solutions of salts instead of the electrolysis of molten salts. **[1 mark]**

! Exam Tip

Look at what is happening at the electrodes in the diagram.

06.2 Explain why test tubes are placed over the electrodes. **[1 mark]**

06.3 **Table 2** shows some of the student's results.

Table 2

Experiment	Electrolyte	Observations at anode	Observations at cathode
1	copper chloride solution	bubbles	
2	copper sulfate solution	bubbles	cathode coated in reddish metal
3	potassium bromide solution	yellow-brown liquid	bubbles

! Exam Tip

Base your predictions on the other results in the table.

Predict what the student would observe at the cathode in experiment **1**. **[1 mark]**

06.4 Describe how to test the gas in the bubbles in experiment **2**. In your answer, give the results you would expect. **[2 marks]**

06.5 Explain how the gas in the bubbles in experiment **3** are formed. Include a half equation in your answer. **[5 marks]**

07 A chemist tried to pass an electric current through a solid, a liquid, and a solution. **Table 3** shows the chemist's results.

! Exam Tip

The gas formed is not sulfur or sulfur dioxide. There are only four gases you are expected to know how to test for in this exam; make sure you know them.

Table 3

Substance	State	Observations at anode	Observations at cathode
sodium chloride	solid	no change (did not conduct electricity)	
sodium chloride	liquid	smell of chlorine	silver-coloured liquid produced
sodium chloride	concentrated solution	gas produced did not relight glowing splint smell of chlorine	gas produced lit splint gives a squeaky pop
sodium chloride	dilute solution	gas produced relit glowing splint smell of chlorine	gas produced lit splint gives a squeaky pop

07.1 Explain the observations in solid and liquid sodium chloride. **[3 marks]**

07.2 Write a half equation, including state symbols, for the reaction that occurs at the cathode for concentrated sodium chloride. **[3 marks]**

07.3 Suggest an explanation for the observations at the anode and cathode for dilute sodium chloride solution. **[6 marks]**

 **Exam Tip**

There are two gases produced here, not one gas that gives two positive results.

08 Aluminium is manufactured by electrolysis.

08.1 Suggest why reduction with carbon is not an appropriate method to manufacture aluminium. **[1 mark]**

08.2 In the electrolysis of aluminium, what is the cathode made of? **[1 mark]**

08.3 A mixture of aluminium oxide and cryolite forms the electrolyte. Explain the purpose of the cryolite. **[3 marks]**

08.4 Explain why aluminium is produced at the cathode. **[2 marks]**

08.5 In the electrolysis of aluminium oxide, explain why the anode has to be replaced regularly. **[2 marks]**

 **Exam Tip**

The electrodes are made of carbon.

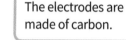

09 A student sets up an electrolysis experiment in a Petri dish, as shown in **Figure 5**.

Figure 5

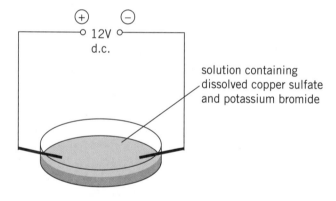

12V d.c.

solution containing dissolved copper sulfate and potassium bromide

Table 4 shows the results.

Table 4

Electrode	Observations
anode	
cathode	brown flaky solid and then bubbles

09.1 Give the name of the brown flaky solid. **[1 mark]**

09.2 Predict the name of the gas that forms bubbles at the cathode. **[1 mark]**

09.3 Describe a test you could do to show that your prediction in **09.2** is correct. Include the expected results of this test. **[2 marks]**

09.4 Predict and explain the observations expected at the anode. Include half equations in your answer. **[8 marks]**

09.5 Suggest **two** reasons for carrying out the electrolysis in a Petri dish, rather than in a larger and taller electrolysis cell. **[2 marks]**

 **Exam Tip**

Use the diagram to determine the ions in the electrolyte.

10 Phosphoric(V) acid, H_3PO_4, is a weak acid.

10.1 What is meant by the term weak acid? **[1 mark]**

10.2 Phosphoric(V) acid can react with magnesium metal:
$$Mg(s) + H_3PO_4(aq) \rightarrow Mg_3(PO_4)_2(aq) + H_2(g)$$
Balance the symbol equation. **[1 mark]**

10.3 Identify which species is oxidised and which species is reduced. **[2 marks]**

10.4 Phosphoric(V) acid can also react with sodium hydroxide.
$$NaOH(aq) + H_3PO_4(aq) \rightarrow \underline{\quad} (aq) + H_2O(l)$$
Give the formula of the salt produced in this reaction, and then balance the equation. **[3 marks]**

10.5 A teacher reacted $25\,cm^3$ of $0.5\,mol/dm^3$ phosphoric(V) acid with a solution of $0.15\,mol/dm^3$ of sodium hydroxide. Calculate the volume of sodium hydroxide solution that reacted with the phosphoric(V) acid. **[5 marks]**

11 In a decomposition reaction, one substance breaks down on heating to form two or more substances.

11.1 Magnesium nitrate decomposes to form solid magnesium oxide, nitrogen dioxide gas, and oxygen gas.

Write a balanced chemical equation for the reaction, including state symbols. **[3 marks]**

11.2 The equation for another decomposition reaction is:
$$CaCO_3(s) \rightarrow CaO(s) + CO_2(g)$$
Give the name of the compound with the formula $CaCO_3$. **[1 mark]**

11.3 Calculate the relative formula mass of $CaCO_3$. **[2 marks]**

11.4 $50\,kg$ of $CaCO_3$ is heated until the decomposition reaction is complete. The reaction produces $22\,kg$ of CO_2. Calculate the mass of CaO that is produced. **[2 marks]**

12 A student has a mixture of coloured liquids. The student wants to identify which substances are in the mixture.

12.1 The student thinks that there are two substances dissolved in water. They carry out a chromatography experiment to identify whether this is accurate. They use water as the solvent. Sketch the chromatogram the student would see if there are two substances dissolved in water. **[3 marks]**

12.2 The student finds that the mixture only contains one substance dissolved in water. They carry out another chromatography experiment to identify the substances. Their chromatogram is shown in **Figure 6**.

Figure 6

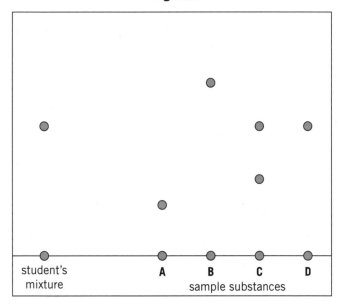

Identify the substance in the student's mixture. **[1 mark]**

> **! Exam Tip**
>
> Use a pen to draw horizontal lines for each spot. This will help you identify which spots appear in more than one sample.

12.3 The student wants to separate the substance from the water it is dissolved in. The substance has a boiling point of 78 °C. Explain why fractional distillation has to be used to extract the substance from the water. **[5 marks]**

13 This question is about the elements in Group 1 and Group 7 of the Periodic Table.

13.1 Describe the pattern in the melting points of the Group 7 elements, from the top to the bottom of the group. **[1 mark]**

13.2 Compare the patterns in the reactivity of the Group 1 and Group 7 elements, from the top to the bottom of the groups. **[2 marks]**

13.3 Name the products formed when sodium reacts with water. **[2 marks]**

> **! Exam Tip**
>
> Make sure it's clear which group you're talking about in each part of your answer.

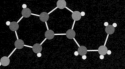

Energy changes

During a chemical reaction, energy transfers occur. Energy can be transferred:

- to the surroundings – **exothermic**
- from the surroundings – **endothermic**

This energy transfer can cause a temperature change. Energy is always conserved in chemical reactions.

This means that there is the same amount of energy in the Universe at the start of a chemical reaction as at the end of the chemical reaction.

The surroundings

When chemists say energy is transferred from or to "the *surroundings*" they mean "everything that isn't the reaction".

For example, imagine you have a reaction mixture in a test tube. If you measure the temperature in the test tube using a thermometer, the thermometer is then part of the surroundings.

- If the thermometer records an increase in temperature, the reaction in the test tube is exothermic.
- If the thermometer records a decrease in temperature, the reaction in the test tube is endothermic.

Reaction profiles

A **reaction profile** shows whether a reaction is exothermic or endothermic.

The **activation energy** is the minimum amount of energy that particles must have to react when they collide.

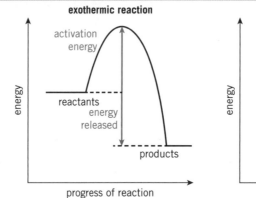
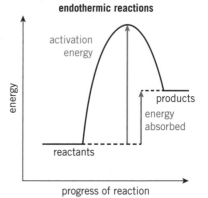

Bonds

Atoms are held together by strong chemical bonds. In a reaction, those bonds are broken and new ones are made between different atoms.

- Breaking a bond requires energy so is endothermic.
- Making a bond releases energy so is exothermic.

Breaking bonds

If a lot of energy is released when making the bonds and only a little energy is required to break them, then overall energy is released and the reaction as a whole is exothermic.

Making bonds

If a little energy is released when making the bonds and a lot is required to break them, then overall energy is taken in and the reaction as a whole is endothermic.

Bond calculations

Different bonds require different amounts of energy to be broken (their **bond energies**). To work out the overall energy change of a reaction, you need to:

 Revision tip

Drawing everything out helps to count the correct number of bonds.

1 work out how much energy is required to break all the bonds in the reactants
2 work out how much energy is released when making all the bonds in the products.

overall energy transferred = energy required to break bonds – energy required to make bonds

- A positive number means an endothermic reaction.
- A negative number means an exothermic number.

A summary of exothermic and endothermic reactions is given in the table.

Reaction	Energy transfer	Temperature change	Example	Everyday use	Bonds
exothermic	to the surroundings	temperature of the surroundings increases	• **oxidation** • **combustion** • **neutralisation**	• self-heating cans • hand warmers	more energy released when making bonds than required to break bonds
endothermic	from the surroundings	temperature of the surroundings decreases	• **thermal decomposition** • citric acid and sodium hydrogen carbonate	• sports injury packs	less energy released when making bonds than required to break bonds

Chemical cells

In a metal displacement reaction, one metal is oxidised – it loses electrons. These electrons are transferred to another metal, which gains the electrons and so is reduced.

By using a **chemical cell** to conduct this reaction, the electron's movement generates a current.

In the cell shown, the zinc atoms from the electrode lose electrons, turn into ions, and move into the solution.

The electrons travel through the circuit to the copper electrode, causing the LED to light up.

Once at the copper electrode, a metal ion *from the solution* will pick the electrons up and become a metal atom.

The greater the difference in reactivity between the two metals in the cell, the greater the potential difference produced.

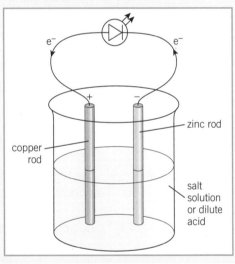

Batteries

A **battery** is formed of two or more cells connected in series.

• Some batteries are **rechargeable**. An external electric current is applied, which reverses the reaction.

• Some batteries, like alkaline batteries, are not rechargeable because the reaction is not reversible. Once the reactants are used up, the chemical reaction stops and no more potential differences are released.

Hydrogen fuel cells

Fuel cells use a fuel and oxygen from the air to generate a potential difference.

Hydrogen fuel cells generate electricity from hydrogen and oxygen. The overall reaction is:

$$2H_2(g) + O_2(g) \rightarrow 2H_2O(l)$$

The hydrogen is oxidised to produce water.

There are different types of hydrogen fuel cell. In alkaline fuel cells, the half equations are below:

• $2H_2(g) + 4OH^-(aq) \rightarrow 4H_2O(l) + 4e^-$
• $O_2(g) + 2H_2O(l) + 4e^- \rightarrow 4OH^-(aq)$

Advantages
• the only waste is water
• do not need to be electrically recharged

Disadvantages
• hydrogen is highly flammable and difficult to store
• hydrogen is often produced from non-renewable resources

 Key terms

Make sure you can write a definition for these key terms.

activation energy   battery

bond energy   chemical cell

combustion   endothermic

exothermic   fuel cell

neutralisation   oxidation

reaction profile   rechargeable

thermal decomposition

 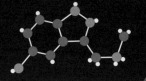 

Retrieval

Learn the answers to the questions below then cover the answers column with
a piece of paper and write as many as you can. Check and repeat.

	C11 questions		Answers
1	What is an exothermic energy transfer?	*Put paper here*	transfer to the surroundings
2	What is an endothermic energy transfer?	*Put paper here*	transfer from the surroundings
3	What is a reaction profile?		diagram showing how the energy changes in a reaction
4	What is the activation energy?	*Put paper here*	minimum amount of energy required before a collision will result in a reaction
5	What is bond energy?		the energy required to break a bond or the energy released when a bond is formed
6	In terms of bond breaking and making, what is an exothermic reaction?	*Put paper here*	less energy is required to break the bonds than is released when making the bonds
7	In terms of bond breaking and making, what is an endothermic reaction?	*Put paper here*	more energy is required to break the bonds than is released when making the bonds
8	How are chemical cells made?		connect two different metals (electrodes) in a solution (electrolyte)
9	What is a battery?		two or more chemical cells connected in series
10	How does the potential difference of a cell depend on the metals that the electrodes are made of?	*Put paper here*	the bigger the difference in reactivity, the greater the potential difference
11	How can some cells be recharged?		by applying an external current
12	Why can some cells not be recharged?	*Put paper here*	the reaction cannot be reversed
13	What is a fuel cell?	*Put paper here*	cell that uses a fuel and oxygen to generate electricity
14	In the hydrogen fuel cell, what is the overall reaction?	*Put paper here*	$2H_2(g) + O_2(g) \rightarrow 2H_2O(l)$
15	In the alkaline hydrogen fuel cells, what are the half equations?	*Put paper here*	$2H_2(g) + 4OH^-(aq) \rightarrow 4H_2O(l) + 4e^-$ $O_2(g) + 2H_2O(l) + 4e^- \rightarrow 4OH^-(aq)$
16	Give an advantage of the hydrogen fuel cell.	*Put paper here*	only product is water, do not need to be electrically recharged
17	Give a disadvantage of the hydrogen fuel cell.	*Put paper here*	hydrogen is flammable, difficult to store and is often produced from non-renewable sources

Now go back and use the questions below to check your knowledge from previous chapters.

C11

Previous questions

Answers

1	What is the half equation for the ionisation of water?	$H_2O\,(l) \rightarrow H^+\,(aq) + OH^-\,(aq)$
2	Why can metals like potassium and aluminium not be extracted by reduction with carbon?	they are more reactive than carbon
3	What is the conservation of mass?	in a chemical reaction, atoms are not created or destroyed, just rearranged, so total mass before = total mass after the reaction
4	Why can solid ionic substances not be electrolysed?	they do not conduct electricity, or the ions cannot move
5	What are concordant titres?	titres within $0.1\,cm^3$ of each other
6	What is a base?	substance that reacts with acids in neutralisation reactions
7	What type of substance is given as ions in an ionic equation?	ionic compounds in solution (or liquid)
8	In the electrolysis of aluminium oxide, why do the anodes need to be replaced?	they react with the oxygen being formed
9	What is an alkali?	substance that dissolves in water to form a solution above pH 7

Put paper here (repeated in centre column)

Required Practical Skills

Practise answering questions on the required practicals using the example below.
You need to be able to apply your skills and knowledge to other practicals too.

Temperature changes	Worked Example	Practice
This practical tests your ability to accurately measure mass, temperature, and volume to investigate changes during chemical reactions. You should be able to describe how to measure temperature change after mixing a strong acid with a strong alkali. You also need to know: • general equations for reactions of acids • how to determine the formula of ionic compounds from the charges of their ions • formulas of ions involved in neutralisation reactions.	Write a method to investigate how the volume of sodium hydroxide added to hydrochloric acid affects the temperature change of the reaction. **Answer:** Place a polystyrene cup inside a beaker. Pour $30\,cm^3$ of dilute hydrochloric acid into the cup, place the lid on the cup and insert a thermometer through a hole in the cup. Record the temperature of the acid. Pour $5\,cm^3$ sodium hydroxide solution into the cup and stir the solution. Record the highest temperature the reaction reaches on the thermometer. Repeat the experiment, increasing the amount of sodium hydroxide each time by $5\,cm^3$, up to $40\,cm^3$. Repeat the entire experiment two times to get repeat measurements.	1 Describe the function of the beaker in the experiment. 2 Is the reaction in this experiment exothermic or endothermic? Explain your answer. 3 Give a balanced equation for the reaction between sodium hydroxide and hydrochloric acid.

01 This question is about endothermic and exothermic reactions.

01.1 Draw **one** line from each reaction to show whether it is endothermic or exothermic. **[4 marks]**

Reaction	Endothermic or exothermic
thermal decomposition	
citric acid with sodium hydrogencarbonate	endothermic
neutralisation	exothermic
combustion	

> **! Exam Tip**
>
> Read the instructions carefully; you may only see the word ONE in bold and think you need to draw ONE line but read the whole thing and you'll see that you need FOUR lines, one from each box on the left-hand side.

01.2 Which statement is true for an exothermic reaction?

Tick **one** box. **[1 mark]**

It transfers energy to the surroundings. ☐

It transfers energy from the surroundings. ☐

The energy of the products is higher than the energy of the reactants. ☐

The temperature of the surroundings decreases. ☐

01.3 Iron oxide reacts with aluminium to produce aluminium oxide and iron. The reaction occurs at a high enough temperature that the iron produced is molten.

Identify whether the reaction is exothermic or endothermic.
[1 mark]

_____

Paper 1 | C11

02 Hydrogen reacts with chlorine to form hydrochloric acid. The reaction is exothermic.

Figure 1 shows the reaction profile for the reaction.

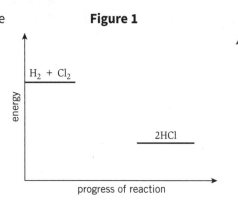

Figure 1

02.1 Complete the reaction profile to show how the energy changes as the reaction proceeds. Draw an arrow to show the overall energy change of the reaction.

The arrow and line do **not** need to be to scale. **[2 marks]**

02.2 Explain how the reaction profile in **Figure 1** shows that the reaction is exothermic. **[2 marks]**

_____

_____

_____

> **! Exam Tip**
>
> You need to refer to the position of the products and reactants in your answer.

02.3 The displayed formulae for the reaction are:

$$H—H + Cl—Cl \rightarrow 2\ H—Cl$$

Table 1 shows some bond enthalpies.

Table 1

	H—H	Cl—Cl	H—Cl
Energy in kJ/mol	436	242	431

Calculate the overall energy change of the reaction. **[3 marks]**

> **! Exam Tip**
>
> Don't forget there are 2 H—Cl bonds, even though only one has been drawn.

Bond energy _____ kJ/mol

03 Some students dissolved four substances in water. This is the method used:

1 Transfer 100 cm³ of water to a beaker.

2 Measure the temperature of the water.

3 Add 1 spatula measure of the substance, in powder form, to the water.

4 Measure the new temperature of the water.

> **! Exam Tip**
>
> This is a Required Practical. You may have done it by mixing acid and alkali; try to remember what you put the acid into. Even though you use a beaker this is not where the reaction occurs.

Table 2 shows their results.

Table 2

Substance	Temperature of water at start in °C	Temperature of solution immediately after dissolving in °C
A	20	25
B	21	17
C	21	31
D	22	6

03.1 Suggest the apparatus that could be used instead of a beaker to improve step **1**. Give a reason for your suggestion. **[2 marks]**

03.2 Suggest what the students should do between steps **3** and **4**. Give a reason for your suggestion. **[2 marks]**

03.3 Give the letter of the substance that dissolves most exothermically. **[1 mark]**

03.4 Predict how the temperature changes would alter if the students repeated the experiment using 200 cm³ of water in step **1**. **[1 mark]**

04 Scientists are looking at using hydrogen fuel cells instead of rechargeable batteries.

04.1 Write the balanced symbol equation for the reaction that occurs in a hydrogen fuel cell. [2 marks]

04.2 Identify which species is oxidised in a hydrogen fuel cell. **[1 mark]**

04.3 Hydrogen fuel cells could be used as an alternative in electric cars, which use rechargeable batteries. Evaluate the use of hydrogen fuel cells compared to a standard electric car battery. **[4 marks]**

05 **Figure 2** shows the displayed formulae for the combustion reaction of methane.

Figure 2

$$\underset{\begin{array}{c}H\\|\\H-C-H(g)\\|\\H\end{array}}{} + 2\,O{=}O(g) \longrightarrow O{=}C{=}O(g) + 2\,\underset{H\quad H}{O}(g)$$

05.1 **Table 3** shows some bond enthalpy values.

Table 3

	C—H	O=O	C=O	O—H
Energy in kJ/mol	412	496	743	463

Figure 3 is the reaction profile for the reaction shown in **Figure 2**. It is **not** drawn to scale.

Calculate the energy change for each arrow **A**, **B**, and **C** shown in **Figure 3**.   [3 marks]

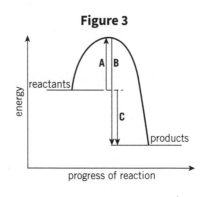

Figure 3

05.2   Name the energy change represented by arrow **C**.   [1 mark]

05.3   Define the energy change represented by arrow **A**.   [1 mark]

06   A student wanted to compare the temperature changes when different metals reacted with hydrochloric acid. The student set up the apparatus shown in **Figure 4**.

Figure 4

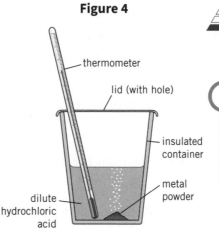

06.1   Name the dependent variable in the investigation.   [1 mark]

06.2   Identify **two** control variables in the investigation.   [2 marks]

> (!) **Exam Tip**
>
> Control variables are all the parts of the experiment that we need to keep the same to make sure it's a fair test.

06.3   The student's results are shown in **Table 4**.

Table 4

Metal	Temperature of acid at start in °C	Temperature of mixture immediately after reaction in °C	Temperature change in °C
magnesium	19.0	36.7	
zinc	19.5	25.6	
copper	20.4	20.4	0.0

Suggest a reason for the result for copper.   [1 mark]

06.4   Complete the missing values in **Table 4**.   [1 mark]

06.5   State which metal reacts most exothermically with hydrochloric acid.   [1 mark]

07   **Figure 5** shows the reaction profiles of four reactions: **A**, **B**, **C**, and **D**.

The reactions profiles are drawn to scale.

Figure 5

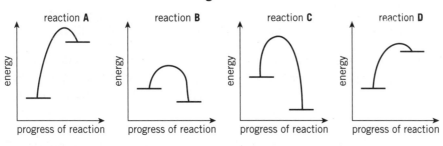

07.1   Give the letters of **two** reaction profiles that could show combustion reactions.   [1 mark]

> (!) **Exam Tip**
>
> To help with this question draw arrows to show the activation energy, from where the reaction starts to its highest point.

07.2 Give the letter of the **one** reaction with the smallest activation energy. **[1 mark]**

07.3 Give the letter of the **one** reaction that is most exothermic. **[1 mark]**

08 Some students wanted to investigate the energy changes of the reactions of hydrochloric acid with three metal carbonates.

08.1 Write a balanced equation for the reaction of copper carbonate with hydrochloric acid. Include state symbols. **[3 marks]**

08.2 Describe a method that the students could use to carry out the investigation. In your answer name the apparatus required and identify the independent, dependent, and control variables. **[6 marks]**

08.3 State how the investigation results show which metal carbonate reacts most exothermically with acid. **[1 mark]**

09 A student makes an electrical cell by connecting two different metals with a salt solution.

09.1 Which metal would produce the largest potential difference when connected with copper? Tick **one** box.

iron ☐   lead ☐   tin ☐

Explain your answer. **[2 marks]**

09.2 Suggest why potassium is not a suitable metal to use. **[1 mark]**

09.3 The student uses zinc as metal **A**. Write an ionic equation for the oxidation reaction that will happen in this chemical cell. **[3 marks]**

10 This question is about two fuels. **Table 5** gives the energy released on burning 1 mole of each of the fuels.

Table 5

Fuel name	Chemical formula of fuel	Energy transferred to the surroundings per mole of fuel burnt in kJ/mol
methane	CH_4	890
nonane	C_9H_{20}	6125

10.1 The equation for the combustion of methane is:

$$CH_4 + 2O_2 \rightarrow CO_2 + 2H_2O$$

Identify which substance is oxidised in the reaction. **[1 mark]**

10.2 The combustion products of nonane are the same as those for methane. Write a balanced chemical equation for the combustion of nonane. **[3 marks]**

10.3 Compare the environmental impacts of burning the two fuels in terms of the energy transferred per gram of fuel burnt and the energy transferred per gram of carbon dioxide produced. Relative atomic masses A_r: C = 12; H = 1; O = 16 **[6 marks]**

11 This question is about strong and weak acids. **Table 6** shows the pH values of four acid solutions.

Table 6

Solution	Acid	Concentration in mol/dm³	pH
W	ethanoic acid	1.0	2.4
X	hydrochloric acid		2.4
Y	ethanoic acid		4.4
Z	chloroethanoic acid	1.0	1.4

11.1 Explain what you can deduce about the concentrations of solutions **X** and **Y** from the data in **Table 6**. **[6 marks]**

Exam Tip

This is just asking for general comments on the concentration; you don't need to do any calculations.

11.2 Use data from **Table 6** to compare the degree of ionisation of chloroethanoic acid in aqueous solution with the degree of ionisation of ethanoic acid in aqueous solution. **[2 marks]**

11.3 Write an ionic equation for the reaction that occurs when an acid is neutralised by an alkali. **[1 mark]**

12 This question is about the reactivity series of metals.

12.1 Compare the reactions of potassium, lithium, and magnesium with cold water. **[6 marks]**

12.2 Caesium is at the bottom of Group 1 of the Periodic Table. Predict the names of the products of the reaction of caesium with water. **[1 mark]**

Exam Tip

Base your answer on what you know about sodium or potassium.

12.3 Explain how the reactivities of the Group 1 elements are related to the tendency of the elements to form their positive ions. **[2 marks]**

13 This question is about the production of the metal lead. Methods **A** and **B** describe two processes for extracting lead.

Method A

1 Dig lead sulfide from the ground.

2 Heat the lead sulfide in air: $2PbS + 3O_2 \rightarrow 2PbO + 2SO_2$

3 Heat the lead oxide with carbon: $2PbO + C \rightarrow 2Pb + CO_2$

Method B

1 Collect battery paste from used batteries.

The paste is a mixture of lead sulfate and lead oxides.

2 React the paste with an alkaline solution.

One product is a soluble sulfate solution.

3 Heat the remaining mixture with carbon, for example:

$$2PbO + C \rightarrow 2Pb + CO_2$$

13.1 Name the substance that is reduced in step **3** of method **A**. **[1 mark]**

13.2 Write a balanced chemical equation for step **2** of method **B**.

• The reactants are lead sulfate and sodium hydroxide.

• The products are lead hydroxide and sodium sulfate. **[3 marks]**

Exam Tip

This is a short two-mark question so you don't need a detailed explanation.

13.3 Suggest **two** advantages of method **B** compared to method **A**. **[2 marks]**

⚙ Knowledge

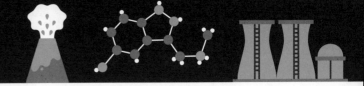

C12 Rate of reaction

Rates of reaction

The **rate of a reaction** is how quickly the reactants turn into the products.

To calculate the rate of a reaction, you can measure:

- how quickly a reactant is used up

$$\text{mean rate of reaction} = \frac{\text{quantity of reactant used}}{\text{time taken}}$$

- how quickly a product is produced.

$$\text{mean rate of reaction} = \frac{\text{quantity of product formed}}{\text{time taken}}$$

For reactions that involve a gas, this can be done by measuring how the mass of the reaction changes or the volume of gas given off by the reaction.

Volume of gas produced

The reaction mixture is connected to a gas syringe or an upside down measuring cylinder. As the reaction proceeds the gas is collected.

The rate for the reaction is then:

$$\text{rate} = \frac{\text{volume of gas produced}}{\text{time taken}}$$

Volume is measured in cm^3 and time in seconds, so the unit for rate is cm^3/s.

Mean rate between two points in time

To get the mean rate of reaction between two points in time:

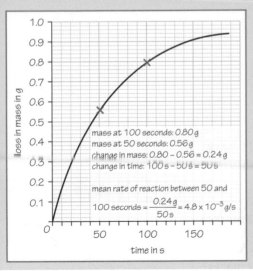

mass at 100 seconds: 0.80 g
mass at 50 seconds: 0.56 g
change in mass: 0.80 – 0.56 = 0.24 g
change in time: 100 s – 50 s = 50 s

mean rate of reaction between 50 and 100 seconds = $\frac{0.24\,g}{50\,s}$ = 4.8 × 10⁻³ g/s

Change in mass

The reaction mixture is placed on a mass balance. As the reaction proceeds and the gaseous product is given off, the mass of the flask will decrease.

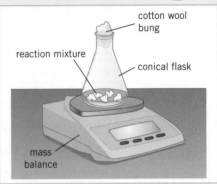

cotton wool bung
reaction mixture
conical flask
mass balance

The rate for the reaction is then:

$$\text{rate} = \frac{\text{change in the mass}}{\text{time taken}}$$

The mass is measured in grams and time is measured in seconds. Therefore, the unit of rate is g/s.

Calculating rate from graphs

The results from an experiment can be plotted on a graph.

- A steep gradient means a high rate of reaction – the reaction happens quickly.
- A shallow gradient means a low rate of reaction – the reaction happens slowly.

Mean rate at specific time

To obtain the rate at a specific time draw a **tangent** to the graph and calculate its **gradient**.

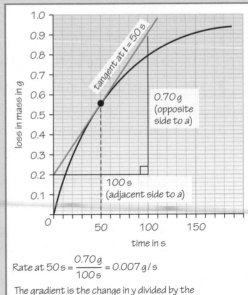

tangent at t = 50 s
0.70 g (opposite side to a)
100 s (adjacent side to a)

Rate at 50 s = $\frac{0.70\,g}{100\,s}$ = 0.007 g/s

The gradient is the change in y divided by the change in x for a right-angled triangle drawn from the tangent.

Collision theory

For a reaction to occur, the reactant particles need to collide. When the particles collide, they need to have enough energy to react or they will just bounce apart. This amount of energy is called the **activation energy**.

You can increase the rate of a reaction by:

- increasing the **frequency of collisions**
- increasing the energy of the particles when they collide.

Factors affecting rate of reaction

Condition that increases rate	How is this condition caused?	Why it has that effect
increasing the temperature	Heat the container in which the reaction is taking place.	1 particles move faster, leading to more frequent collisions 2 particles have more energy, so more collisions result in a reaction note that these are two *separate* effects
increasing the concentration of solutions	Use a solution with more solute in the same volume of solvent.	there are more reactant particles in the reaction mixture, so collisions become more frequent
increasing the pressure of gases	Increase the number of gas particles you have in the container or make the container smaller.	less space between particles means more frequent collisions
increasing the surface area of solids	Cut the solid into smaller pieces, or grind it to create a powder, increasing the surface area. Larger pieces decrease the surface area.	only reactant particles on the surface of a solid are able to collide and react; the greater the surface area the more reactant particles are exposed, leading to more frequent collisions

Catalysts

Some reactions have specific substances called **catalysts** that can be added to increase the rate. These substances are not used up in the reaction.

A catalyst provides a different reaction pathway that has a lower activation energy. As such, more particles will collide with enough energy to react, so more collisions result in a reaction.

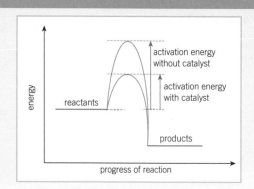

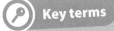 **Key terms**

Make sure you can write a definition for these key terms.

activation energy   catalyst   collision   collision theory   frequency of collision

gradient   rate of reaction   tangent

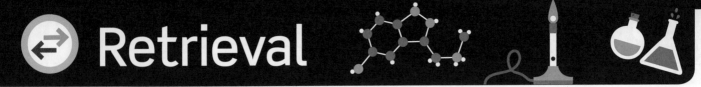

Retrieval

Learn the answers to the questions below then cover the answers column with a piece of paper and write down as many as you can. Check and repeat.

C12 questions

Answers

	C12 questions		Answers
1	What is the rate of a reaction?	Put paper here	how quickly reactants are used up or products are produced
2	What is the equation for calculating the mean rate of reaction?		$\text{mean rate} = \dfrac{\text{change in quantity of product or reactant}}{\text{time taken}}$
3	What is the unit for rate of reaction in a reaction involving a change in mass?		g/s
4	What is the unit for rate of reaction in a reaction involving a change in volume?	Put paper here	cm^3/s
5	What is the activation energy?		the minimum amount of energy colliding particles have to have before a reaction will take place
6	What effect does increasing concentration have on the rate of reaction?		increases
7	Why does increasing concentration have this effect?		more reactant particles in the same volume lead to more frequent collisions
8	What effect does increasing pressure have on the rate of reaction?	Put paper here	increases
9	Why does increasing pressure have this effect?		less space between particles means more frequent collisions
10	What effect does increasing surface area have on the rate of reaction?	Put paper here	increases
11	Why does increasing surface area have this effect?		more reactant particles are exposed and able to collide, leading to more frequent collisions
12	What effect does increasing temperature have on the rate of reaction?		increases
13	Why does increasing temperature have this effect?	Put paper here	particles move faster, leading to more frequent collisions – particles have the same activation energy, so more collisions result in a reaction
14	What is a catalyst?		a substance that increases the rate of a reaction but is not used up in the reaction
15	How do catalysts increase the rate of a reaction?		lower the activation energy of the reaction, so more collisions result in a reaction

Now go back and use the questions below to check your knowledge from previous chapters.

C12

Previous questions

Answers

Put paper here

1 Give a disadvantage of the hydrogen fuel cell.

hydrogen is flammable, difficult to store and is often produced from non-renewable sources

2 What charges do ions from Groups 1 and 2 form?

Group 1 forms 1+, Group 2 forms 2+

3 How does the hardness of a Transition Metal compare to that of a Group 1 metal?

higher (Transition Metals)

4 In the alkaline hydrogen fuel cells, what are the half equations?

$2H_2 (g) + 4OH^- (aq) \rightarrow 4H_2O(l) + 4e^-$ $O_2(g) + 2H_2O(l) + 4e^- \rightarrow 4OH^-(aq)$

5 In terms of H+ ions, what is an acid?

a substance that releases H^+ ions when dissolved in water

6 What are the products of a reaction between a metal hydroxide and an acid?

salt + water

7 How does the potential difference of a cell depend on the metals that the electrodes are made of?

the bigger the difference in reactivity, the greater the potential difference

8 In the electrolysis of a metal halide solution, what is produced at the anode?

halogen

9 In the hydrogen fuel cell, what is the overall reaction?

$2H_2(g) + O_2(g) \rightarrow 2H_2O(l)$

Required Practical Skills

Practise answering questions on the required practicals using the example below. You need to be able to apply your skills and knowledge to other practicals too.

Rates of reaction	Worked Example	Practice
From this practical, you should be able to describe two ways in which the rate of a reaction can be measured. These are: **1** measuring the production of a gas **2** measuring changes in the colour or turbidity of a solution You need to be able to describe the method for collecting gas with an inverted measuring cylinder, and for measuring the colour or turbidity change in a reaction. There are different methods of measuring rates of reaction, but remember that these principles are applicable to all of them.	Silver chloride is an insoluble salt that can be made in the following reaction. Suggest a how the rate of this reaction could be measured. $AgNO_3(aq) + NaCl(aq) \rightarrow NaNO_3(aq) + AgCl(s)$ **Answer:** The reactants are both colourless solutions. Solid silver chloride will form as a precipitate and make the solution appear cloudy. One way of measuring the rate of the reaction is to look at the rate of production of silver chloride precipitate. This could be measured by placing the beaker with the reacting solution on a piece of white paper with a black cross printed on it, and measuring the time taken for the cross to disappear.	**1** Give three factors that can affect the rate of a reaction. **2** Give two methods that can be used to determine the rate of a reaction where a gas is produced. **3** Suggest another method to measure the rate of the production of silver chloride precipitate.

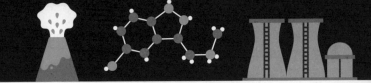

Practice

Exam-style questions

01 Potassium iodide acts as a catalyst for the decomposition of hydrogen peroxide.

01.1 What does a catalyst do? **[1 mark]**

Tick **one** box.

Decreases reaction rate by providing a pathway with a higher activation energy. ☐

Decreases reaction rate by providing a pathway with a lower activation energy. ☐

Increases reaction rate by providing a pathway with a higher activation energy. ☐

Increases reaction rate by providing a pathway with a lower activation energy. ☐

> **! Exam Tip**
>
> The are two parts to **01.1**. The first part relates to reaction rate and the second part relates to activation energy. Decide how a catalyst affects reaction rate and cross off the two wrong answers. Then decide how a catalyst affects the activation energy and cross of the one remaining wrong answer.

01.2 **Figure 1** shows the reaction profile for the catalysed reaction.

Figure 1

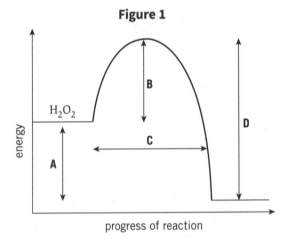

Which arrow shows the activation energy for the catalysed reaction? **[1 mark]**

_____

01.3 Balance the symbol equation for the decomposition of hydrogen peroxide. Give the missing state symbol. **[2 marks]**

_____H_2O_2(aq) → _____H_2O(_____) + O_2(g)

01.4 Why is potassium iodide not given in the balanced symbol equation? **[1 mark]**

_____

02   Magnesium reacts with dilute hydrochloric acid:

$$Mg + 2HCl \rightarrow MgCl_2 + H_2$$

A student investigated how the volume of hydrogen produced changed over time.

Table 1 shows the student's results.

Table 1

Time in seconds	Total volume of hydrogen produced in cm³
0	0
30	22
60	38
90	52
120	58
150	61
180	61

Figure 2

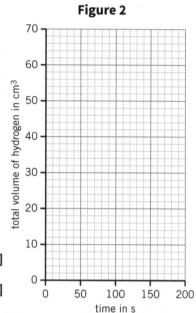

02.1   Plot the data on **Figure 2**. [2 marks]

02.2   Draw a line of best fit.   [1 mark]

02.3   Calculate the rate of reaction at 100 seconds.

Give your answer to two significant figures.   [4 marks]

Rate at 100 seconds = _____ cm³/s

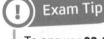

02.4   The rate of the reaction at 10 seconds is 0.83 cm³/s.

Suggest a reason for the difference in rate at 10 seconds and 100 seconds.   [1 mark]

_____

_____

> **Exam Tip**
>
> It might help you if you mark 30, 60, 90 etc. on the x-axis to help you accurately plot the points.

> **Exam Tip**
>
> To answer **02.3** you need to use a tangent to the curve at 100 seconds. Draw the largest triangle you can fit on the graph to help you determine the rate of reaction.

03 Sodium thiosulfate solution reacts with hydrochloric acid. One of the products of the reaction is sulfur, which forms as a precipitate.

$$Na_2S_2O_3(aq) + 2HCl(aq) \rightarrow 2NaCl(aq) + H_2O(l) + SO_2(g) + S(s)$$

Some students investigated the rate of reaction at different temperatures. This is the method used:

1 Place 50 cm³ of sodium thiosulfate in a conical flask.

2 Start a timer then add 5 cm³ hydrochloric acid to the flask.

3 Look down through the flask at a cross drawn on a piece of paper.

4 Stop the timer when the cross disappears.

5 Repeat the experiment, each time heating the sodium thiosulfate to a different temperature.

03.1 Suggest **one** improvement that could be made to step **2** to ensure that the results at different temperatures are comparable. **[1 mark]**

03.2 One student suggests measuring the temperature of the reaction mixture after adding the hydrochloric acid instead of measuring the temperature of the sodium thiosulfate on its own. Suggest an advantage of this idea. **[1 mark]**

03.3 Suggest why the same student in the group should carry out step **4** at every temperature. **[1 mark]**

03.4 **Table 2** shows the students' results. Identify which result is anomalous. **[1 mark]**

03.5 Describe how the rate of reaction changes between 0 °C and 61 °C. **[1 mark]**

03.6 Explain why the rate of reaction changes between 0 °C and 61 °C. **[2 marks]**

Table 2

Temperature in °C	Time for X to disappear in seconds
0	180
21	44
39	22
45	21
52	9
61	5

04 Calcium carbonate reacts with hydrochloric acid:

$$2HCl(aq) + CaCO_3(s) \rightarrow CaCl_2(aq) + CO_2(g) + H_2O(l)$$

Some students want to investigate how the size of the pieces of solid calcium carbonate affects the rate of the reaction. **Figure 3** shows the apparatus.

Figure 3

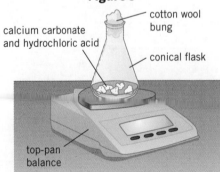

calcium carbonate and hydrochloric acid

cotton wool bung

conical flask

top-pan balance

This is the method used:

1 Weigh approximately 3.0 g of calcium carbonate onto a piece of paper and leave it all on the balance.

2 Place 100 cm³ of dilute hydrochloric acid in a conical flask and place on the balance.

3 Zero the balance.

4 Add the calcium carbonate to the acid and start the stopwatch.

5 Leave the flask and its contents on the balance but remove the paper.

6 Record the time for the total mass to decrease by 0.50 g.

7 Repeat with different sized pieces of calcium carbonate.

04.1 Explain why the total mass of the contents of the conical flask decreases. **[1 mark]**

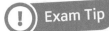

Exam Tip

Look at the state symbols of the products.

04.2 Suggest an improvement to step **5**. Give a reason for your answer. **[2 marks]**

04.3 **Table 3** shows the students' results.

Exam Tip

The students have used the same mass of calcium carbonate (as stated in the method). What will be the difference between one large piece of calcium carbonate and lots of small pieces of calcium carbonate?

Table 3

Size of calcium carbonate pieces	Time for mass to decrease by 0.50 g in seconds
large	1280
medium	690
small	302

Explain the pattern shown in **Table 3**.

[3 marks]

05 Zinc is a metal. It reacts with dilute nitric acid. The products of the reaction are zinc nitrate and a gas.

05.1 Name the gas formed in the reaction. **[1 mark]**

05.2 Which one of these changes will make the reaction rate **slower**? Choose **one** answer. **[1 mark]**

decreasing the acid concentration

increasing the pressure

decreasing the size of the pieces of zinc

increasing the temperature

Exam Tip

05.2 is asking about temperature and rate in the opposite way to its normally asked. The chemistry is still the same so just apply your knowledge to the slightly different situation.

05.3 Predict the effect of decreasing the temperature on the rate of reaction. **[1 mark]**

06 Some students investigated the factors that affect the rate of the reaction of magnesium with excess hydrochloric acid. They followed the reaction by measuring the total volume of gas formed every 30 seconds. They changed the conditions and repeated the experiment. **Figure 4** is a graph of some of the results of the two experiments, **P** and **Q**. Curve **P** shows the results for the first experiment.

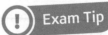

Figure 4

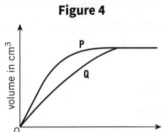

! **Exam Tip**

Both reactions end at the same point, thats why the graphs levels off at the same point for both lines. The rate will only affect how *quickly* a reaction finishes, not how much product is produced.

06.1 Write the balanced symbol equation with state symbols for the reaction between magnesium and hydrochloric acid. **[3 marks]**

06.2 Curve **Q** was obtained in the second experiment. Suggest **one** variable that the students might have changed and **two** variables that they kept constant in order to obtain curve **Q**. Justify your answer. **[6 marks]**

07 A student investigated the catalytic decomposition of hydrogen peroxide:

$$2H_2O_2\ (aq) \rightarrow 2H_2O(l)\ +\ O_2(g)$$

Figure 5 shows the apparatus used.

Figure 5

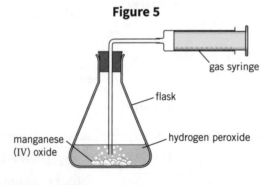

gas syringe

flask

manganese (IV) oxide

hydrogen peroxide

07.1 The student made a mistake in setting up the apparatus. Describe how the student must improve the apparatus before doing the experiment. Give a reason for making this improvement. **[2 marks]**

07.2 The student improved the apparatus set-up and collected some data. **Table 4** shows the results the student obtained.

Calculate the average rate of reaction. Give the unit of the rate of the reaction.

[3 marks]

Table 4

Time in minutes	Volume of gas produced in cm³
0	0
1	42
2	69
3	86
4	88
5	91
6	91

! **Exam Tip**

The number of significant figures in your answer should match the data provided.

! **Exam Tip**

Use the units of the variables to help you work out the units for the rate.

07.3 Predict how the mean rate of reaction would change if powdered manganese(IV) oxide was used instead of lumps. Give a reason for your prediction. **[2 marks]**

08 A student investigated the decomposition of magnesium carbonate.

$$MgCO_3(s) \rightarrow MgO(s) + CO_2(g)$$

The student measured the volume of carbon dioxide produced. **Table 6** shows their results.

Table 6

Time in s	0	20	40	60	80	100	120	140	160
Volume of carbon dioxide produced in cm³	0	20	39	55	64	65	73	75	75

08.1 Plot the student's results on **Figure 6**. Draw a line of best fit. **[3 marks]**

Figure 6

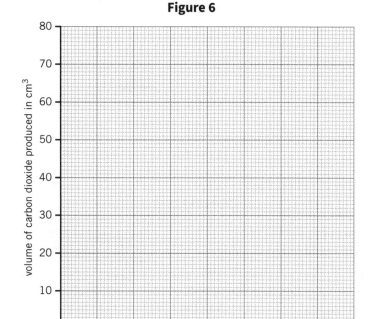

> **! Exam Tip**
>
> Your points must be plotted using a sharp pencil, a cross placed directly over the point you're indicating. Dots or circles are not clear enough as the can cover many points.

> **! Exam Tip**
>
> Your line of best fit may be straight or curved and must go thought the majority of the points, but not always all of them. Draw it with a smooth single line.

08.2 Identify the anomalous point. **[1 mark]**

08.3 Explain, in terms of particles, why the rate of reaction was faster at the start of the reaction than towards the end of the reaction. **[4 marks]**

08.4 Use **Figure 6** to calculate the rate of reaction at 60 seconds. **[2 marks]**

09 A student investigates the effect of concentration of the rate of reaction between nitric acid and sodium carbonate.

09.1 Write a method the student could use. Your method should include how you will measure the rate of reaction and the variables you will control. **[6 marks]**

09.2 Write a prediction for the student's investigation. Explain your prediction. **[3 marks]**

09.3 Another student investigated the effect of the surface area of sodium carbonate on the rate of the reaction. Their results are shown in **Table 5**.

Table 5

Surface area of sodium carbonate	Time taken to produce 500 cm³ of carbon dioxide in s	Mean rate of reaction in _____
solid pieces	195	2.7
powder	42	

Give the unit of the rate of reaction. **[1 mark]**

09.4 Calculate the mean rate of reaction for powdered sodium carbonate. Give your answer to three significant figures. **[2 marks]**

09.3 Give **one** other factor that will affect the rate of the reaction between nitric acid and sodium carbonate. **[1 mark]**

10 Sodium thiosulfate reacts with hydrochloric acid in the following reaction:

$$Na_2S_2O_3(aq) + 2HCl(aq) \rightarrow 2NaCl(aq) + H_2O(l) + SO_2(g) + S(s)$$

10.1 Explain how increasing the concentration of hydrochloric acid will affect the rate of the reaction. **[1 mark]**

10.2 Explain how increasing the temperature will affect the rate of the reaction. **[4 marks]**

10.3 The rate of the reaction between sodium thiosulfate and hydrochloric acid can be determined by measuring how long it takes for the reaction mixture to become cloudy. A student uses a light sensor and data logger to measure the turbidity cloudiness of the reaction. Sketch a graph of time against turbidity to show the rate of the reaction at two different temperatures. **[4 marks]**

! Exam Tip

Sketching graphs doesn't mean plotting points, just the labelled axis and the line are all thats needed.

11 A scientist sets up a hydrogen fuel cell with an alkaline electrolyte.

11.1 Write the half equation that occurs at the negative electrode. Include state symbols. **[3 marks]**

11.2 Write the half equation that occurs at the positive electrode. Include state symbols. **[3 marks]**

11.3 Evaluate the use of hydrogen fuel cells compared to rechargeable cells. **[4 marks]**

12 Lead nitrate solution reacts with sodium iodide solution to make lead iodide, PbI_2, and sodium nitrate, $NaNO_3$. The lead iodide forms as a precipitate. Sodium nitrate is in solution.

12.1 Define the law of the conservation of mass. **[2 marks]**

12.2 Lead nitrate reacts with sodium iodide. Write a balanced equation, including state symbols, for the reaction. The formula of lead nitrate is $Pb(NO_3)_2$. **[3 marks]**

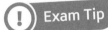
12.3 Calculate the minimum mass of sodium iodide required to make 6.68 g of lead iodide. Give your answer to two significant figures. Relative atomic mass A_r: Na = 23; Pb = 207; I = 127 **[5 marks]**

13 This question is about the elements in Groups 7 and 0 of the Periodic Table.

13.1 Give the electron configurations of fluorine and neon. Use the Periodic Table to help you to answer this question. **[2 marks]**

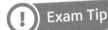
13.2 Explain why the Group 7 elements are very reactive but the Group 0 elements are unreactive. **[2 marks]**

13.3 Describe the pattern in the boiling points of the Group 7 and Group 0 elements. **[1 mark]**

14 **Figure 7** shows one form of carbon. Its formula is C_{60}.

Figure 7

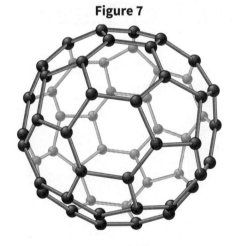

14.1 Give the name of the form of carbon shown in **Figure 7**. **[1 mark]**

14.2 Compare the bonding and structure of graphite with the form of carbon shown in **Figure 7**. **[3 marks]**

14.3 Explain **two** properties of diamond. **[4 marks]**

 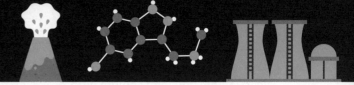

Knowledge

C13 Equilibrium

Reversible reactions

In some reactions, the products can react to produce the original reactants. This is called a **reversible reaction**. When writing chemical equations for reversible reactions, use the $\rightleftharpoons$ symbol.

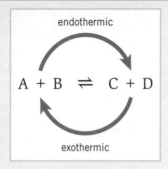

endothermic

$$A + B \rightleftharpoons C + D$$

exothermic

In this reaction:

- A and B can react to form C and D – the forward reaction
- C and D can react to form A and B – the reverse reaction.

The different directions of the reaction have opposite energy changes.

If the forward reaction is *endothermic*, the reverse reaction will be *exothermic*.

The same amount of energy is transferred in each direction.

Equilibrium

In a **closed system** no reactants or products can escape. If a reversible reaction is carried out in a closed system, it will eventually reach **dynamic equilibrium** – a point in time when the forward and reverse reactions have the same rate.

At dynamic equilibrium:

- the reactants are still turning into the products
- the products are still turning back into the reactants
- *the rates of* these two processes are *equal*, so overall the amount of reactants and products are constant.

Dynamic equilibrium

At dynamic equilibrium the amount of reactant and product are constant, but not necessarily equal.

You could have a mixture of reactants and products in a 50:50 ratio, in a 75:25 ratio, or in any ratio at all. The **conditions** of the reaction are what change that ratio.

How dynamic equilibrium is reached

Progress of reaction	start of reaction	middle of reaction	at dynamic equilibrium
Amount of A + B	high	decreasing	constant
Frequency of collisions A + B	high	decreasing	constant
Rate of forward reaction	high	decreasing	same as rate of reverse reaction

forward reaction

equilibrium is reached at this point

rate of reaction

reverse reaction

time

Amount of C + D	zero	increasing	constant
Frequency of collisions C + D	no collisions	increasing	constant
Rate of reverse reaction	zero	increasing	same as rate of forward reaction

C13

Reaction conditions

The conditions of a reaction refer to the external environment of the reaction. When the reaction occurs in a closed system, you can change the conditions by:

- changing the concentration of one of the substances
- changing the temperature of the entire reaction vessel
- changing the pressure inside the vessel.

Le Châtelier's principle

At equilibrium, the amount of reactants and products is constant. In order to change the amounts of reactant and product at equilibrium the *conditions* of the reaction must be changed. The closed system will then counteract the change by favouring either the forward reaction or the reverse reaction. This is known as **Le Châtelier's principle**.

For example, lowering the concentration of the product in the system causes the forward reaction to be **favoured** to increase the concentration of the product.

Changing concentrations

Change	Effect	Explanation
decrease concentration of product	favours the forward reaction	opposes the change by making *less* reactant and *more* product
increase concentration of product	favours the reverse reaction	opposes the change by making *more* reactant and *less* product
decrease concentration of reactant	favours the reverse reaction	opposes the change by making *more* reactant and *less* product
increase concentration of reactant	favours the forward reaction	opposes the change by making *less* reactant and *more* product

Changing temperature

The effect of changing the temperature depends on which direction is exothermic and which direction is endothermic.

This varies from reaction to reaction – some are exothermic in the forward direction, but others will be exothermic in the reverse direction.

Change	Effect	Explanation
increase temperature of surroundings	favours the endothermic reaction	opposes the change by decreasing the temperature of the surroundings
decrease temperature of surroundings	favours the exothermic reaction	opposes the change by increasing the temperature of the surroundings

Changing pressure

The effect of changing the pressure depends on which side of the reaction has more molecules of gas (e.g., the reaction $2A(g) + B(g) \rightleftharpoons A_2B(g)$ has three molecules of gas on the reactant side and one molecule on the product side). If both sides have the same number of molecules, then changing the pressure will have no effect.

Change	Effect	Explanation
increase the pressure	favours the reaction that results in fewer molecules	decreasing the number of molecules within the vessel opposes the change because it decrease pressure
decrease the pressure	favours the direction that results in more molecules	increasing the number of molecules within the vessel opposes the change because it increase pressure

Key terms

Make sure you can write a definition for these key terms.

closed system   conditions   dynamic equilibrium   Le Châtelier's principle   reversible reaction

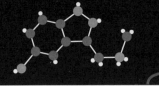

Learn the answers to the questions below then cover the answers column with
a piece of paper and write down as many as you can. Check and repeat.

C13 questions | Answers

	Question	Answer
1	What is a reversible reaction?	the reactants turn into products and the products turn into reactants
2	Which symbol shows a reversible reaction?	$\rightleftharpoons$
3	What is dynamic equilibrium?	the point in a reversible reaction when the rate of the forward and reverse reactions are the same
4	What are the three reaction conditions that can be changed?	concentration, temperature, pressure
5	What is Le Châtelier's principle?	the position of equilibrium will shift to oppose external changes
6	What is the effect of increasing the concentration of reactants on a reaction at dynamic equilibrium?	favours the forward reaction
7	What is the effect of decreasing the concentration of products on a reaction at dynamic equilibrium?	favours the forward reaction
8	What is the effect of increasing pressure on a reaction at dynamic equilibrium?	favours the reaction that leads to the fewest molecules
9	What is the effect of decreasing pressure on a reaction at dynamic equilibrium?	favours the reaction that leads to the most molecules
10	What is the effect of increasing temperature on a reaction at dynamic equilibrium?	favours the endothermic reaction
11	What is the effect of decreasing temperature on a reaction at dynamic equilibrium?	favours the exothermic reaction

Put paper here

Now go back and use the questions below to check your knowledge from previous chapters.

Previous questions

	Previous questions	Answers
1	What is the equation for calculating the mean rate of reaction?	mean rate = change in quantity of product or reactant / time taken
2	How do the densities of Transition Metals compare to those of Group 1 metals?	higher (Transition Metals)
3	How do catalysts increase the rate of a reaction?	lower the activation energy of the reaction, so more collisions result in a reaction
4	How are chemical cells made?	connect two different metals (electrodes) in a solution (electrolyte)
5	Give the distinctive properties of Transition Metal compounds.	form ions with different charges and coloured compounds
6	What is the activation energy?	the minimum amount of energy colliding particles have to have before a reaction will take place
7	Why are the noble gases inert?	they have full outer shells so do not need to lose or gain electrons
8	In the electrolysis of aluminium oxide, why is the aluminium oxide mixed with cryolite?	to lower the melting point
9	What is an endothermic energy transfer?	transfer from the surroundings

Put paper here

Put paper here

Put paper here

Maths Skills

Practise your maths skills using the worked example and practice questions below.

Ratios, fractions, and percentages

In chemistry we often use ratios, fractions, and percentages to describe mixtures. These are different mathematical forms of numbers that represent the same thing.

A **ratio** compares the size of two or more quantities.

A **fraction** can express a part of a whole number, or shows one number divided by another in an equation.

A **percentage** is a number expressed as a fraction of 100.

Worked Example

A stoppered flask contains a gas that is a mixture of 70 atoms of neon and 50 atoms of helium.

The **ratio** of neon atoms to helium atoms in the mixture is 70:50, which simplifies to 7:5 by dividing each side by the highest common factor (in this case, 10).

The **fraction** of atoms that are neon is:

$$\frac{70}{(70+50)} = \frac{70}{120} = \frac{7}{12}$$

The **percentage** of atoms that are helium is:

$$\left(\frac{50}{(70+50)}\right) \times 100 = 41.67\%$$

Practice

1. An equilibrium mixture contains 45 cm^3 of H_2 and 22.5 cm^3 of O_2. What is the ratio of H_2 to O_2?
2. What fraction of the total volume in this mixture is O_2?
3. A different equilibrium mixture contains 92 cm^2 of H_2, 154 cm^2 of N_2, and 23 cm^3 of NH_3. What is the ratio of the three different substances?
4. What is the percentage of NH_3 and N_2 combined out of the whole mixture?

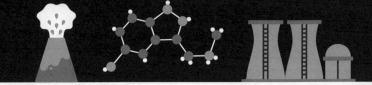

Practice

Exam-style questions

01 Ammonia is formed from the reversible reaction between nitrogen and hydrogen.

$$N_2(g) + 3H_2(g) \rightleftharpoons 2NH_3(g)$$

01.1 The forward reaction transfers 92 kJ of energy to the surroundings. State how much energy is transferred by the reverse reaction. **[1 mark]**

_____

> **! Exam Tip**
>
> Energy is always conserved.

01.2 Define Le Châtelier's Principle. **[1 mark]**

_____

_____

_____

01.3 The reacting mixture is placed in apparatus that prevents the escape of reactants and products.

The pressure of the reaction mixture is then increased.

What happens to the position of the equilibrium?

Tick **one** box. **[1 mark]**

It does not change. ☐

It shifts to the left. ☐

> **! Exam Tip**
>
> Look at the number of moles of each side of the reaction.

It shifts to the right. ☐

It shifts to the left and then to the right. ☐

02 Sulfur dioxide reacts with oxygen in a reversible reaction.

$$2SO_2(g) + O_2(g) \rightleftharpoons 2SO_3(g)$$

The forward reaction is exothermic.

02.1 Draw a dot and cross diagram to show the bonding in an oxygen molecule, O_2. **[2 marks]**

02.2 Define the term exothermic. **[1 mark]**

_____

_____

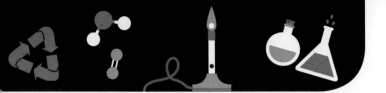

Paper 2   C13

02.3   The reversible reaction above occurs in a closed container.
Give the effect on the position of the equilibrium for each of the
following condition changes.   **[4 marks]**

more SO_3 is added: _____

pressure is increased: _____

temperature is increased: _____

more O_2 is added: _____

Exam Tip

The equilibrium position shift to counter the change; this means the reaction will go in the opposite direction to the change made.

03   Methanol is used a fuel. It can be produced by reacting carbon
monoxide with hydrogen in a reversible reaction.

$$CO(g) + 2H_2(g) \rightleftharpoons CH_3OH(g)$$

The forward reaction is exothermic.

03.1   Explain why this reaction can only reach equilibrium in a
sealed container.   **[3 marks]**

03.2   Identify **three** factors that will affect the position of the equilibrium
reaction to produce methanol. Explain the effect that changing each
of these factors has on the position of the equilibrium.   **[6 marks]**

03.3   Calculate the maximum mass of methanol that can be made from
10.0 g of carbon monoxide. Give your answer to three significant
figures. Use the Periodic Table to help you.   **[6 marks]**

Exam Tip

This is a 6 mark question asking for three factors; you will get one mark for each factor and one mark for explaining its effect on the equilibrium.

04   Hydrogen reacts with iodine to form hydrogen iodide.

$$H_2(g) + I_2(g) \rightleftharpoons 2HI(g)$$

04.1   State what does the $\rightleftharpoons$ symbol tells you about the reaction.   **[1 mark]**

04.2   The reaction reaches equilibrium in apparatus that prevents the
escape of reactants and products. Describe what happens to the
particles of H_2, I_2, and HI at equilibrium.   **[1 mark]**

04.3   The forward reaction is endothermic. Describe the energy transfers
involved in the forward reaction.   **[2 marks]**

Exam Tip

Use the correct keywords to get the marks in this answer.

05   Some students want to investigate the reversible change:
hydrated copper sulfate $\rightleftharpoons$ anhydrous copper sulfate + water

Figure 1 shows the apparatus.

Figure 1

05.1   Explain why equilibrium cannot
be reached using the apparatus in
Figure 1.   **[1 mark]**

05.2   Suggest a suitable piece of
equipment for heating the
hydrated copper sulfate. **[1 mark]**

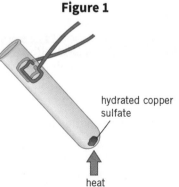

hydrated copper
sulfate

heat

05.3 Name the substance that leaves the test tube. **[1 mark]**

05.4 Suggest how the substance that leaves the test tube could be collected. **[3 marks]**

 Exam Tip

You only need to suggest how to collect it, not test it.

06 Some students investigated the equilibrium reaction between two solutions containing cobalt ions:

pink cobalt ion solution + chloride ions ⇌ blue cobalt ion solution + water

Figure 2 shows the apparatus.

06.1 Suggest an improvement to the apparatus to make sure that all the cobalt ion solution is at the same temperature. **[1 mark]**

Figure 2

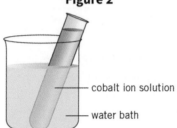

— cobalt ion solution

— water bath

06.2 The students carry out some tests. **Table 1** shows their results.

Table 1

Test number	Action	Initial colour change
1	heat the water bath	from pink to blue
2	add ice to the water bath	from blue to pink
3	add chloride ions to the solution	
4	add water to the solution	

Write a conclusion based on the data in rows **1** and **2** in **Table 1**. **[2 marks]**

06.3 Predict the colour change that would be observed in row **3** of **Table 1**. **[1 mark]**

06.4 Explain the prediction you made to answer **06.3**. **[3 marks]**

 Exam Tip

Remember this is a reversible reaction, so the reaction will change direction to counter the change made in the environment.

07 A student sets up an equilibrium of two nitrogen oxides in a sealed gas syringe. The equilibrium is represented by the equation:

$$N_2O_4(g) \rightleftharpoons 2NO_2(g)$$

colourless   brown

 **Exam Tip**

Look at the number of moles on each side of the equation.

At equilibrium the substances in the syringe are light brown in colour. The student recorded their observations when they moved the plunger of the gas syringe (**Table 2**).

Table 2

Action	Colour change
push in the syringe plunger	light brown to colourless
pull out the syringe plunger	light brown to dark brown

07.1 Explain what happens to the equilibrium when the plunger is pushed in. **[4 marks]**

Exam Tip

Don't use abbreviations or short hand in your answer; in class you might be used to LHS for left-hand side but in the exam you need to write out the words in full.

07.2 The forward reaction is endothermic. The student places the syringe in an ice–water mixture. Predict and explain the colour change observed. **[3 marks]**

07.3 The student sets another equilibrium up in a separate gas syringe:
$$H_2(g) + I_2(g) \rightleftharpoons 2HI(g)$$

Predict and explain the effect on the position of equilibrium of increasing the pressure on the equilibrium mixture. **[2 marks]**

08 The reaction between two substances, **X** and **Y**, is reversible:
$$X(g) \rightleftharpoons Y(g)$$

Substance **Y** is placed in a sealed container. After some time, equilibrium is established. **Figure 3** shows how the concentrations of **X** and **Y** change as equilibrium is established.

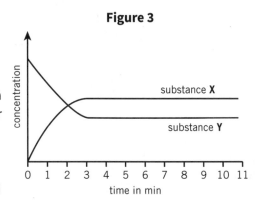

Figure 3

08.1 Identify the time at which the equilibrium was established. Give a reason for your answer. **[2 marks]**

08.2 State how the rate of the forward and reverse reactions compares at equilibrium. **[1 mark]**

08.3 The system was heated for 5 minutes. At the end of the 5 minutes, it was found that there was more of substance **X** in the system than before the system was heated. Identify which reaction is exothermic. **[1 mark]**

09 A teacher had a closed system at equilibrium that contained Cl_2 (a pale green gas), ICl_3 (yellow crystals), and ICl (a brown liquid). The teacher removed Cl_2 from the system, and left the system to reach equilibrium. At the new equilibrium, the amount of ICl had increased and the amount of ICl_3 had decreased.

09.1 Write a balanced symbol equation with state symbols for the reaction. **[3 marks]**

09.2 The teacher placed the system into an ice bath. More yellow crystals formed. Identify whether the formation of ICl_3 is exothermic or endothermic. **[1 mark]**

09.3 Explain why placing the system in an ice bath favours this reaction. **[3 marks]**

10   A student heats a sample of ammonium chloride in a test tube. The ammonium chloride breaks down into ammonia gas and hydrogen chloride gas.

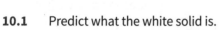
solid ammonium chloride $\rightleftharpoons$ ammonia gas + hydrogen chloride gas

As the student heats the ammonium chloride, a white solid forms at the top of the test tube (**Figure 4**).

Figure 4

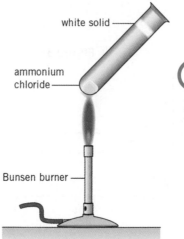

white solid

ammonium chloride

Bunsen burner

10.1   Predict what the white solid is.

[1 mark]

! **Exam Tip**

Look for the clues in the main body of the question.

10.2   Explain your prediction in **10.1**.

[3 marks]

10.3   Another student heats hydrated blue copper(II) sulfate crystals in a test tube. Water is given off to form white anhydrous copper(II) sulfate crystals. This process is reversible. Predict what would happen if water was added to the anhydrous copper sulfate crystals.

[1 mark]

10.4   The student holds the test tube whilst they add the water. Predict what they would feel as water is added.

[1 mark]

11   Ammonia, NH_3, is a gas at room temperature and pressure.

11.1   Give the number of moles of hydrogen atoms in one mole of ammonia gas.

[1 mark]

11.2   Calculate the relative molecular mass of ammonia. **[2 marks]**

! **Exam Tip**

You'll need to look up the mass numbers in the Periodic Table.

11.3   Calculate the number of moles of ammonia in a 68 g sample of the gas.

[2 marks]

11.4   Calculate the number of particles of ammonia in a 68 g sample of the gas. The Avogadro constant is 6.02×10^{23} per mole. Give your answer to three significant figures.

[3 marks]

12   Some metals react with water.

12.1   Describe the expected observations in the reaction of potassium and water.

[2 marks]

12.2   Name the products of the reaction of lithium with water. **[2 marks]**

! **Exam Tip**

An observation is what you see happening.

12.3   Describe the expected observations when copper is placed in a test tube of water.

[1 mark]

13 A student wanted to make sodium chloride crystals from sodium hydroxide and hydrochloric acid solutions:

$$NaOH(aq) + HCl(aq) \rightarrow NaCl(aq) + H_2O(l)$$

The student used the following method:

1 Use a measuring cylinder to transfer 25 cm³ of sodium hydroxide into a conical flask.

2 Add 1 cm³ of indicator.

3 Add dilute hydrochloric acid from a burette until the indicator changes colour.

4 Pour the mixture from the flask into a beaker.

5 Heat the beaker and its contents until half the water has evaporated.

6 Allow the rest of the water to evaporate by leaving the beaker in a warm, dry place.

13.1 Suggest and explain **one** improvement to step **1** and **one** improvement to step **2**. **[4 marks]**

13.2 Describe what the students should do between steps **3** and **4**. Give a reason for this extra step. **[2 marks]**

13.3 The students start with 0.025 mol of sodium hydroxide. Calculate the maximum mass of sodium chloride they can expect to make. Give your answer to two significant figures. **[4 marks]**

> **! Exam Tip**
>
> Can you think of a better way to measure these?

> **! Exam Tip**
>
> If you do not give your answer to two significant figures, then you won't get full marks. The correct resolution is an important part of the answer.

14 Sulfuric acid is produced in a multi-step process called the contact process. One step within the contact process involves the reaction between sulfur dioxide and oxygen.

14.1 Complete and balance the symbol equation for the reaction between sulfur dioxide and oxygen. **[3 marks]**

$$\underline{\hspace{1cm}}(g) + O_2(\underline{\hspace{1cm}}) \rightleftharpoons \underline{\hspace{1cm}}SO_3(g)$$

14.2 Increasing the temperature of the reaction vessel causes the equilibrium position to shift to the left. Identify which reaction is exothermic. **[1 mark]**

14.3 What does this suggest about the energy transfers involved in the chemical reaction? **[1 mark]**

14.4 In industry, the reaction is carried out at 450 °C rather than room temperature. Suggest what effect this will have on the yield of sulfur trioxide. **[1 mark]**

14.5 Suggest why the reaction is carried out at 450 °C. **[1 mark]**

14.6 Vanadium pentoxide is used as a catalyst for the forward reaction. Explain how the catalyst increases the rate of the forward reaction. **[3 marks]**

14.7 Suggest and explain **one** other condition that would favour the formation of sulfur trioxide. **[3 marks]**

14.8 Suggest a reason why this condition may not be used in the industrial process. **[1 mark]**

> **! Exam Tip**
>
> Only write in the gaps! Do not try to add in extra numbers or letters outside of the gaps.

 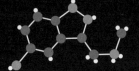 

Knowledge

C14 Crude oil and fuels

Crude oil

Crude oil is incredibly important to our society and economy. It is formed from the remains of ancient biomass – living organisms (mostly plankton) that died many millions of years ago.

Raw crude oil is a thick black liquid made of a large number of different compounds mixed together. Most of the compounds are **hydrocarbons** of various sizes. Hydrocarbons are molecules made of carbon and hydrogen only.

Combustion

Hydrocarbons are used as **fuels**. This is because when they react with oxygen they release a lot of energy. This reaction is called **combustion**. Complete combustion is a type of combustion where the only products are carbon dioxide and water.

Properties

Whether or not a particular hydrocarbon is useful as a fuel depends on its properties:

- **flammability** – how easily it burns
- **boiling point** – the temperature at which it boils
- **viscosity** – how thick it is

Its properties in turn depend on the length of the molecule.

Chain length	Flammability	Boiling point	Viscosity
long chain	low	high	high (very thick)
short chain	high	low	low (very runny)

Alkanes

One family of hydrocarbon molecules are called **alkanes**. Alkane molecules only have single bonds in them. The first four alkanes are:

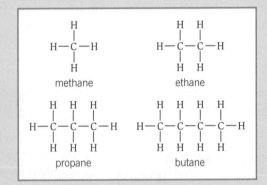

The different alkanes have different numbers of carbon atoms and hydrogen atoms. You can always work the molecular formula of an alkane by using C_nH_{2n+2}.

 Revision tip

You can check if you've drawn compounds correctly since carbon always forms four bonds and hydrogen always forms one bond.

 **Key terms**

Make sure you can write a definition for these key terms.

alkanes   alkenes   boiling point   combustion   cracking   crude oil   feedstock
flammability   fractional distillation   fuel   hydrocarbon   viscosity

Fractional distillation

The different hydrocarbons in crude oil are separated into fractions based on their boiling points in a process called **fractional distillation**. All the molecules in a fraction have a similar number of carbon atoms, and so a similar boiling point.

The process takes place in a fractionating column, which is hot at the bottom and cooler at the top.

The process works like this:

1 crude oil is vapourised (turned into a gas by heating)
2 the hydrocarbon gases enter the column
3 the hydrocarbon gases rise up the column
4 as hydrocarbon gases rise up the column they cool down
5 when the different hydrocarbons reach their boiling point in the column they condense
6 the hydrocarbon fraction is collected.

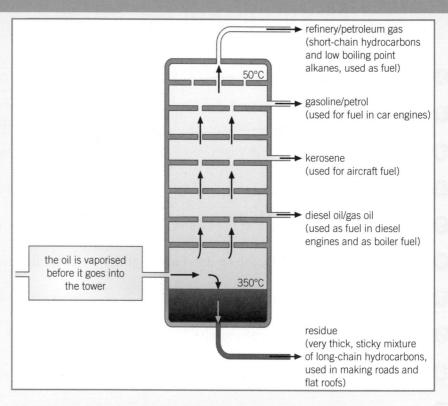

Products from fractional distillation

Many useful products come from the separation of crude oil by fractional distillation.

Fuels	Feedstock	Useful materials produced
petrol, diesel oil, kerosene, heavy fuel oil, and liquefied petroleum gases	fractions form the raw material for other processes and the production of other substances	solvents, lubricants, polymers, and detergents

Cracking

Not all hydrocarbons are as useful as each other. Longer molecules tend to be less useful than shorter ones. As such, there is a higher demand for shorter-chain hydrocarbons than longer-chain hydrocarbons.

A process called **cracking** is used to break up longer hydrocarbons and turn them into shorter ones.

Cracking produces shorter alkanes and **alkenes**.

Two methods of cracking are:

• catalytic cracking – vaporise the hydrocarbons, then pass them over a hot catalyst
• steam cracking – mix the hydrocarbons with steam at a very high temperature

Alkenes

Alkenes are a family of hydrocarbons that contain double bonds between carbon atoms.

Alkenes are also used as fuels, and to produce polymers and many other materials.

They are much more reactive than alkanes. When mixed with bromine water, the bromine water turns from orange to colourless. This can be used to tell the difference between alkanes and alkenes.

Retrieval

Learn the answers to the questions below then cover the answers column with a piece of paper and write down as many as you can. Check and repeat.

	C14 questions	Answers
1	What is a hydrocarbon?	compound containing carbon and hydrogen only
2	How is crude oil formed?	over millions of years from the remains of ancient biomass
3	What are the alkanes?	hydrocarbons that only have single bonds
4	What are the first four alkanes?	methane, ethane, propane, butane
5	What is the general formula for the alkanes?	C_nH_{2n+2}
6	How does boiling point depend on the chain length?	longer the chain, higher the boiling point
7	How does viscosity depend on chain length?	longer the chain, higher the viscosity
8	How does flammability depend on chain length?	longer the chain, lower the flammability
9	How can the different alkanes in crude oil be separated?	fractional distillation
10	What is a fraction?	a group of hydrocarbons with similar chain lengths
11	Name five useful fuels produced from fractional distillation.	petrol, diesel oil, kerosene, heavy fuel oil, and liquefied petroleum gases
12	Name four useful materials produced from crude oil fractions.	solvents, lubricants, polymers, detergents
13	What is cracking?	breaking down a hydrocarbon with a long chain into smaller molecules
14	Name two methods to carry out cracking.	steam cracking and catalytic cracking
15	What are the products of cracking?	short chain alkanes and alkenes
16	What are alkenes?	hydrocarbons with a double bond
17	What are alkenes used for?	formation of polymers
18	Describe the reactivity of alkenes compared to alkanes.	alkenes are much more reactive
19	How can you test for alkenes?	alkenes turn orange bromine water colourless

Put paper here

Now go back and use the questions below to check your knowledge from previous chapters.

C14

Previous questions

Answers

1. What is the effect of decreasing pressure on a reaction at dynamic equilibrium?

favours the reaction that leads to the most molecules

2. In terms of pH, what is an acid?

a solution with a pH of less than 7

3. In terms of oxygen, what is oxidation?

addition of oxygen

4. What is an electrolyte?

liquid or solution that contains ions and so can conduct electricity

5. What is an exothermic energy transfer?

transfer to the surroundings

6. What is the effect of decreasing temperature on a reaction at dynamic equilibrium?

favours the exothermic reaction

7. What is a catalyst?

a substance that increases the rate of a reaction but is not used up in the reaction

8. What is the rate of a reaction?

how quickly reactants are used up or products are produced

9. What is Le Châtelier's principle?

the position of equilibrium will shift to oppose external changes

Put paper here *Put paper here* *Put paper here* *Put paper here*

Maths Skills

Practise your maths skills using the worked example and practice questions below.

Finding the mean	Worked Example	Practice
Whenever an experiment is conducted, it is important to repeat it to establish how *precise* the values are (how close to each other they are), and how *repeatable* they are (can they be repeated). Whenever you repeat an experiment and record repeat observations you must calculate a mean to give an average result for that observation. However, only use values that are close together, and discard any anomalous values.	A student burns ethanol and uses the heat released to warm up some water. As soon as the water increases by 10 °C, she stops the reaction and measures the mass of ethanol used. She repeats this three more times and records the masses: 5.1 g, 6.3 g, 6.5 g, 6.2 g. Calculate the mean of the values. **Step 1**: Establish which values to use – in this case 6.3, 6.5, and 6.2. The first mass (5.1) is ignored because it is not close to the others. **Step 2**: Calculate the mean. $$\text{mean} = \frac{\text{sum of values}}{\text{total number of values}}$$ $$= \frac{(6.3 + 6.5 + 6.2)}{3} = 6.3\,\text{g}$$	1 A student measures how the mass of a magnesium strip increases when burnt in oxygen. They record the masses: 0.12 g, 0.12 g, 0.14 g, 0.11 g, 0.23 g. Calculate the mean increase in mass. 2 The volume of gas produced in three repeats of an experiment is collected, and recorded as: 54 cm³, 58 cm³, 55 cm³. Calculate the mean volume of gas produced.

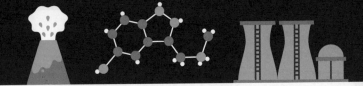
Exam-style questions

01 This question is about alkanes.

01.1 Draw **one** line from each displayed formula to the name of the alkane. **[2 marks]**

Displayed formula

H—C—C—H with H above and below each C (ethane structure)

H—C—C—C—C—H with H above and below each C (butane structure)

Name

| butane |

| ethane |

| methane |

| propane |

01.2 What is the formula of the alkane with 22 carbon atoms?

Tick **one** box. **[1 mark]**

$C_{22}H_{42}$ ☐ $C_{22}H_{46}$ ☐

$C_{22}H_{44}$ ☐ $C_{22}H_{48}$ ☐

> **! Exam Tip**
>
> Use the general formula for alkanes to figure out the number of hydrogen atoms that will be in the compound.

01.3 Decane is an alkane with 10 carbon atoms.

How do the properties of decane compare with the properties of ethane?

Tick **one** box. **[1 mark]**

Decane has a higher flammability, lower boiling point, and higher viscosity. ☐

Decane has a higher flammability, higher boiling point, and lower viscosity. ☐

Decane has a lower flammability, higher boiling point, and higher viscosity. ☐

Decane has a lower flammability, lower boiling point, and lower viscosity. ☐

> **! Exam Tip**
>
> There are three possible differences within each answer; go over the properties one at a time (flammability, boiling point, and then viscosity), comparing them to ethane. This should leave you with the correct answer at the end.

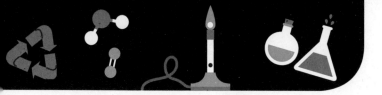

02   **Table 1** shows the boiling points of some alkanes.

Table 1

Name of alkane	Number of carbon atoms	Boiling point in °C
pentane	5	36
hexane	6	69
heptane	7	98
octane	8	126
nonane	9	
decane	10	174
undecane	11	196
dodecane	12	216

02.1   Plot the data from **Table 1** on **Figure 1**.

Draw a line of best fit.   **[3 marks]**

> **! Exam Tip**
>
> Always plot your points with crosses – this shows the examiner exactly which point you are using.

Figure 1

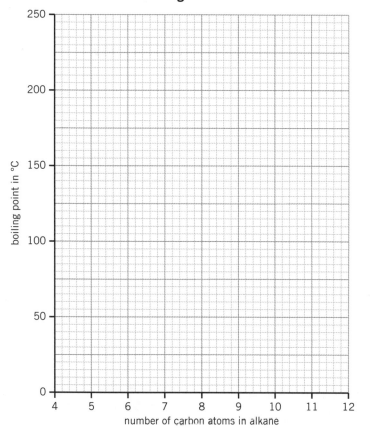

boiling point in °C

number of carbon atoms in alkane

02.2   Use your graph from **Figure 1** to predict the boiling point of nonane.   **[1 mark]**

02.3   Write a chemical equation for the complete combustion of nonane. You do not need to include state symbols.   **[4 marks]**

> **! Exam Tip**
>
> The products for complete combustion are always the same two. The only things that change with large compounds is the number of moles.

03 Many useful substances are obtained from crude oil.

03.1 Define the term 'fraction of crude oil'. **[1 mark]**

03.2 **Figure 2** shows a fractionating column.

Explain how kerosene is separated from the other fractions in crude oil by fractional distillation. **[3 marks]**

Figure 2

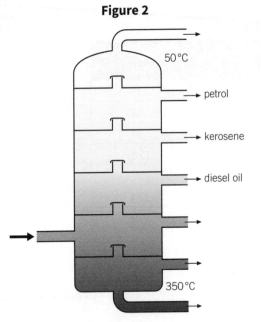

03.3 The hydrocarbons in diesel are bigger than the hydrocarbons in petrol. Compare the physical and chemical properties of diesel and petrol. **[6 marks]**

> **Exam Tip**
>
> Mark on the diagram where it is hottest and where it is coldest; this will point you towards the way they are separated in question **03.2** and the differences in properties in **03.3**.

04 Some students use the apparatus shown in **Figure 3** to crack hydrocarbons.

Figure 3

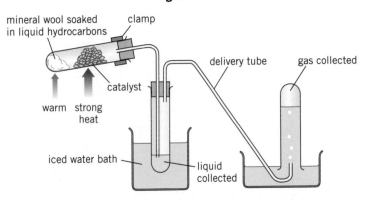

> **Exam Tip**
>
> Think about the changes that would happen if you just heated the liquid hydrocarbons.

04.1 Give a reason for heating the mineral wool soaked in liquid hydrocarbons. **[1 mark]**

04.2 Identify the error in how the students set up the delivery tube. **[1 mark]**

04.3 The equation shows one reaction that occurs in the apparatus.

$$C_{10}H_{22} \rightarrow C_5H_{12} + C_3H_6 + C_2H_4$$

One of the products of the reaction above is collected as a liquid. Write down the formula of the product that you predict is collected as a liquid. Give a reason for your prediction. **[2 marks]**

> **Exam Tip**
>
> Look at the length of the hydrocarbons to help your prediction.

04.4 The product with the formula C_3H_6 is an alkene. Describe the colour change that occurs when C_3H_6 reacts with bromine water. **[1 mark]**

05 A chemist does some tests on four hydrocarbons. The hydrocarbons have the formulae below. The chemist does not know which hydrocarbon is which.

$$C_2H_4 \quad C_2H_6 \quad C_8H_{18} \quad C_{17}H_{36}$$

Table 2 shows the results of the tests.

Table 2

Hydrocarbon	Boiling point in °C	Observations when shaken with bromine water
A	126	no change
B	−104	orange to colourless
C	302	no change
D	−89	no change

05.1 Deduce the formula of each hydrocarbon in **Table 2**. Justify your decision. **[4 marks]**

05.2 Predict the letter of the hydrocarbon in **Table 2** that is most viscous in the liquid state. **[1 mark]**

05.3 In a cracking reaction, a molecule of $C_{20}H_{42}$ forms two compounds:

$$C_8H_{18} \quad C_3H_6$$

Write a balanced chemical equation for the cracking reaction. Do **not** include state symbols. **[3 marks]**

06 This question is about crude oil.

06.1 Name the type of substances that crude oil is made up of. **[1 mark]**

06.2 Describe the process by which crude oil is separated into fractions. **[6 marks]**

06.3 Name **one** of the fractions produced from crude oil, and give its use. **[2 marks]**

07 **Table 3** gives the formulae and boiling points of three alkanes. The alkanes have the same numbers of carbon and hydrogen atoms, but the atoms are joined together differently. Compound **X** is a straight-chain alkane. Compounds **Y** and **Z** are branched-chain alkanes

> **! Exam Tip**
>
> Use the general formula for alkanes and alkenes to first work out which compound belongs to which homologous group.

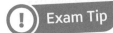

> **! Exam Tip**
>
> Start by trying to find the number of each product by looking at the carbon atoms.

> **! Exam Tip**
>
> The easy examples to remember are the ones that are in cars!

Table 3

	compound **X**	compound **Y**	compound **Z**
Formula	(structure)	(structure)	(structure)
Boiling point in °C	68	63	58

07.1 Name the main source of alkanes. [1 mark]

07.2 Calculate the relative formula mass of the alkanes in **Table 3**.
Relative atomic masses A_r: C = 12; H = 1 [1 mark]

07.3 Explain the relationship between boiling points and the number of branches in a molecule. [6 marks]

 **Exam Tip**

Look at the number of branches the compounds on the previous page have and link the increase in branches to a change in boiling point.

08 This question is about cracking.

08.1 Compare the conditions used for steam cracking and for catalytic cracking. [3 marks]

08.2 The equation shows a cracking reaction:
$$C_{10}H_{22} \rightarrow C_6H_{14} + C_2H_4$$
Balance the equation by writing a number where required. [1 mark]

08.3 Give **two** reasons for carrying out cracking reactions in industry. [2 marks]

 Exam Tip

Start by balancing the carbons; the hydrogens should fall in line behind that

09 Some alkanes are used as fuels.

09.1 Write a balanced equation for the combustion of pentane to make carbon dioxide and water only. [3 marks]

09.2 In certain conditions, pentane undergoes incomplete combustion:
$$2C_5H_{12} + 11O_2 \rightarrow 10CO + 12H_2O$$
Deduce the conditions in which propane undergoes incomplete combustion. Use your answer to **09.1** and the equation above. Justify your answer. [2 marks]

09.3 Propane is another fuel. **Figure 4** shows the displayed formula equation for its complete combustion.

Figure 4

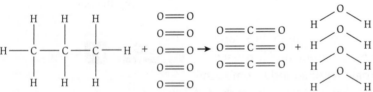

Table 4 gives some bond enthalpy values.

 Exam Tip

Cross off each bond in the diagram as you count it — this should help you count everything only once and not miss bonds out.

Table 4

	C—C	C—H	O = O	C = O	O—H
Energy in kJ/mol	348	412	496	743	463

Calculate the energy change for the complete combustion of 1 mole of propane [5 marks]

10 A teacher passed an electric current through molten zinc chloride. **Figure 5** shows the apparatus.

Figure 5

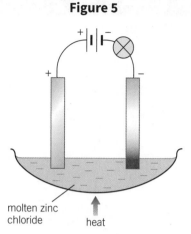

molten zinc chloride

heat

10.1 Predict the observations at the positive and negative electrodes. **[2 marks]**

10.2 Write a half equation for the reaction that occurs at the negative electrode. **[3 marks]**

10.3 The teacher passes an electric current through an aqueous zinc chloride solution. Predict the products formed at the positive and negative electrodes. Zinc is more reactive than hydrogen. **[2 marks]**

11 The equation shows a chemical reaction:

ammonium chloride ⇌ ammonia + hydrogen chloride

11.1 Give the meaning of the symbol ⇌ . **[1 mark]**

11.2 The reaction occurs from left to right if ammonium chloride is heated. State the condition required for the reverse reaction to occur. **[1 mark]**

11.3 State **one** feature of the apparatus required for an equilibrium to be established between ammonium chloride, ammonia, and hydrogen chloride. **[1 mark]**

12 Crude oil is a mixture of many different substances. The substances are separated using fractional distillation.

12.1 Describe what crude oil was formed from. **[1 mark]**

12.2 Name the **two** processes that occur during fractional distillation. **[2 marks]**

12.3 Outline the uses of some products obtained from crude oil. **[6 marks]**

13 Heptane, C_7H_{16} is an alkane.

13.1 Name and describe the bonding between the atoms in a heptane molecule. **[2 marks]**

13.2 Name the type of force that is overcome when heptane boils. **[1 mark]**

13.3 Calculate the mass of carbon dioxide produced when 85.0 g of heptane undergoes complete combustion. Give your answer to two significant figures. **[7 marks]**

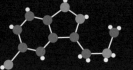
C15 Organic reactions

Homologous series	Functional group	First four of homologous series	Formation	Uses
alkenes	C=C	ethene, C_2H_4; propene, C_3H_6; butene, C_4H_8; pentene, C_5H_{10}	cracking	• formation of polymers • a chemical feedstock
alcohols	–OH	methanol; ethanol; propanol; butanol	Ethanol can be formed from the **fermentation** of sugar – warm a sealed mixture of yeast and a sugar solution. glucose → ethanol + carbon dioxide $C_6H_{12}O_6(aq) \rightarrow 2C_2H_5OH(aq) + 2CO_2(g)$	• *ethanol* is used in alcoholic drinks • first four alcohols mix easily with water, so are used as solvents for substances that don't dissolve in water • common in perfumes, aftershaves and mouthwashes
carboxylic acids	$-C\underset{O-H}{\overset{O}{}}$	methanoic acid; ethanoic acid; propanoic acid; butanoic acid	oxidation of alcohols	• ethanoic acid is used in vinegar

🔑 **Key terms**

Make sure you can write a definition for these key terms.

addition reaction   alcohols   alkene   alkoxide   carboxylic acid   ester   fermentation

There are lots of different 'families' of carbon-containing compounds, for example, alkanes and **alkenes**. These families are called a **homologous series**. Each compound within a homologous series has similar properties and reactions. They all contain specific atoms in specific orders, called the **functional group**.

Combustion reaction	Other reactions	Other information
• complete combustion produces carbon dioxide and water • incomplete combustion more likely, resulting in a smoky yellow flame • both types of alkene combustion release less energy per mole than alkanes	**Addition with halogens** The two atoms from the halogen molecule are *added* across the carbon – carbon double bond. $C_2H_4 + Br_2 \rightarrow C_2H_4Br_2$ **Addition with hydrogen** The two atoms from the hydrogen molecule are *added* across the carbon – carbon double bond to form an alkane. $C_2H_4 + H_2 \rightarrow C_2H_6$ **Addition with steam** React with steam at high temperature and pressure in the presence of a catalyst to form alcohols. $C_2H_4 + H_2O \rightarrow C_2H_5OH$	Alkenes are called **unsaturated** because they have double bonds. As such, atoms can be added to the molecule by breaking the double bond. This contrasts with alkanes which are called **saturated** as there is no space to add more atoms. Alkenes have a general formula C_nH_{2n}.
• short alcohols are very effective fuels and combust easily, burning with a blue flame and producing carbon dioxide and water $2CH_3OH + 3O_2 \rightarrow$ $\quad 2CO_2 + 4H_2O$	**Reaction with sodium** Alcohols react with sodium to release hydrogen. The product from this reaction is called an **alkoxide**, which if added to water forms a strongly alkaline solution. **Oxidation** Alcohols can react with **oxidising agents**, like potassium dichromate, to form carboxylic acids.	
• carboxylic acids can undergo combustion, but we do not generally do this or use them as a fuel	Carboxylic acids react in the same way as other acids. **Reaction with sodium carbonate** Carboxylic acids react with bases to form salts. For example, carboxylic acids react with a metal carbonate to produce a salt, carbon dioxide, and water. **Reaction with alcohols** Carboxylic acids react with alcohols to make water and **esters**. The reaction requires sulfuric acid as a catalyst. Esters have distinctive smells and are used in perfumes and flavourings. The product of ethanol and ethanoic acid is ethyl ethanoate.	When added to water, carboxylic acids are partially ionised to form weakly acidic solutions. They are weak acids.

functional group   homologous series   oxidation   oxidising agent   saturated   unsaturated

Retrieval

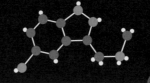

Learn the answers to the questions below then cover the answers column with a piece of paper and write as many as you can. Check and repeat.

	C15 questions	Answers
1	What is a homologous series?	a group of compounds with the same functional group
2	What is a functional group?	a group of atoms that determines the properties of a compound
3	What are alkenes?	a homologous series with a double carbon–carbon bond
4	What is the general formula for alkenes?	C_nH_{2n}
5	What is the product from an addition reaction of an alkene with a halogen?	a haloalkane
6	What is the product from the addition reaction of an alkene with hydrogen?	an alkane
7	What conditions are required for the addition reaction of an alkene with steam?	high temperature, high pressure, and a catalyst
8	What are alcohols?	a homologous series with an –OH group
9	How are alcohols produced?	steam with an alkene or fermentation
10	What conditions are required to produce alcohols by fermenting?	sugar solution with yeast mixed in, warm, sealed vessel
11	Name the first four alcohols.	methanol, ethanol, propanol, butanol
12	What are the products of a reaction between an alcohol and sodium?	hydrogen and an alkoxide
13	What is the organic product formed by the oxidation of an alcohol?	carboxylic acid
14	Name an oxidising agent.	acidified potassium dichromate
15	What are carboxylic acids?	a homologous series with a –COOH group
16	What do carboxylic acids form when they react with sodium carbonate?	salt, carbon dioxide, water
17	How are carboxylic acids produced?	oxidation of alcohols
18	Name the first four carboxylic acids.	methanoic acid, ethanoic acid, propanoic acid, butanoic acid
19	What is the organic product of a reaction between a carboxylic acid and an alcohol?	an ester
20	What catalyst is normally used in the formation of esters?	concentrated sulfuric acid
21	What occurs when pure carboxylic acids are added to water?	a weak acid is formed

Put paper here

Now go back and use the questions below to check your knowledge from previous chapters.

Previous questions | Answers

	Previous questions	Answers
1	What effect does increasing surface area have on the rate of reaction?	increases
2	Why does increasing surface area have this effect?	more reactant particles are exposed and able to collide, leading to more frequent collisions
3	What is a reversible reaction?	the reactants turn into products and the products turn into reactants
4	How can you test for alkenes?	alkenes turn orange bromine water colourless
5	What is the effect of increasing pressure on a reaction at dynamic equilibrium?	favours the reaction that leads to the fewest molecules
6	Name five useful fuels produced from fractional distillation.	petrol, diesel oil, kerosene, heavy fuel oil, and liquefied petroleum gases
7	Which formula is used to calculate volume from concentration and mass?	$\text{volume (dm}^3) = \frac{\text{mass (g)}}{\text{concentration (g/dm}^3)}$
8	How is crude oil formed?	over millions of years from the remains of ancient biomass

Put paper here

Maths Skills

Practise your maths skills using the worked example and practice questions below.

Plotting curves

Remember that you need to draw a line of best fit when numerical data are plotted on a graph.

Some data will need a curved line of best fit, rather than a straight one.

It is important to remember that you draw the line that best fits the data.

Worked Example

The alkanes have the boiling points given below.

Plot the data on a graph, and draw an appropriate line of best fit.

Number of carbon atoms	Boiling point in °C
1	−162.0
2	−89.0
3	−42.0
4	0.0
5	36.0
6	69.0

A graph displaying this data will have a curved line of best fit.

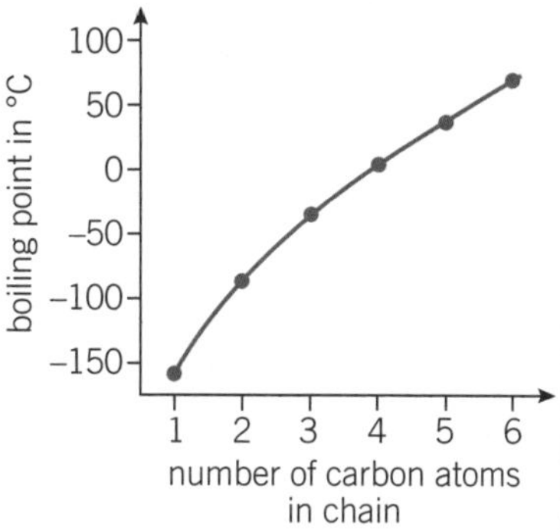

The graph shows a positive correlation – as the number of carbon atoms increases, so does the boiling point.

Practice

A different group of hydrocarbons have the boiling points given below.

Number of carbon atoms	Boiling point in °C
4	5.1
5	44.7
6	72.8
8	106.0
9	112.4
10	115.6

1 Plot a graph of these results. Draw an appropriate line of best fit.

2 Use your graph from **1** to predict the boiling point of a hydrocarbon in this group with seven carbon atoms.

3 Does your graph show a positive or negative correlation?

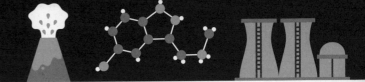
01 **Figure 1** shows the boiling points of some alkanes and alcohols.

Figure 1

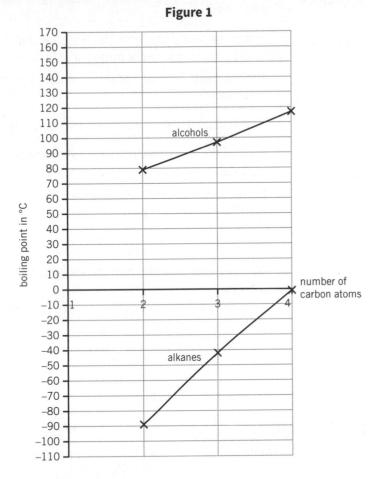

01.1 Name the compound with a boiling point of −42 °C. **[1 mark]**

_____

01.2 Name the compound with a boiling point of 79 °C. **[1 mark]**

_____

01.3 **Table 1** shows the boiling points of some carboxylic acids.

Table 1

Number of carbon atoms	Boiling point in °C
2	118
3	141
4	164

Plot the boiling points of the carboxylic acids on **Figure 1**. Draw a
line of best fit. **[2 marks]**

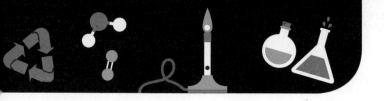
01.4 Describe **two** patterns shown in **Figure 1**.   [2 marks]

1 _____

_____

2 _____

_____

01.5 Explain the pattern shown in the boiling points for the three alcohols.   [2 marks]

_____

_____

_____

02.1 Draw **one** line from the name of each compound to its displayed formula.   [4 marks]

Name   **Displayed formula**

| ethene |

| ethanol |

| butane |

| butanol |

 **Exam Tip**

The are two parts to a hydorcarbon name.

The first part is the number of carbon atoms:

meth-   1 carbon
eth-   2 carbons
prop-   3 carbons
but-   4 carbons

The seconds part that tells us the functional groups:

-ene   has double bonds
-ane   has single bonds
-ol   is an alcohol with an –OH group

02.2 Which homologous series does the compound with the formula CH_3CH_2COOH belong to?

Tick **one** box.   [1 mark]

alcohols ☐   alkenes ☐

alkanes ☐   carboxylic acids ☐

 **Exam Tip**

For both **02.2** and **02.3**, look for the functional group, then count up the number of carbon atoms.

02.3 Write the name of the compound with the formula C_4H_8. **[1 mark]**

_____

03 A student has a test tube of ethanoic acid. The student adds some sodium carbonate powder to the test tube.

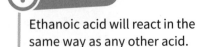

03.1 Explain **one** observation the student would make. **[2 marks]**

Exam Tip

Ethanoic acid will react in the same way as any other acid.

03.2 A teacher mixes some ethanoic acid and ethanol together, then adds a few drops of a catalyst and warms the mixture. A reaction occurs and an ester is formed. Give the purpose of the catalyst. **[1 mark]**

03.3 Name the ester formed in the reaction. **[1 mark]**

03.4 Describe how the ester made can be detected. **[1 mark]**

04 A student bubbles compound **A** through bromine solution.

compound A

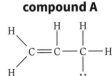

04.1 Describe the observation that the student would make. **[1 mark]**

04.2 Draw the structural formula of the product formed in the reaction of compound **A** with bromine solution. **[1 mark]**

04.3 A new sample of compound **A** is reacted with hydrogen gas. Describe the conditions required for the reaction. **[2 marks]**

04.4 Draw the structural formula of the product formed in the reaction of compound **A** with hydrogen and give its name. **[2 marks]**

Exam Tip

For **04.1** you need to describe a colour change. It is not enough to say the solution turns clear. If you don't use the correct terms, you will not get the mark.

05 The hydrogen atoms in the methyl group of ethanoic acid can be replaced with chlorine atoms. **Table 2** gives the pH of some of these acids.

Table 2

Acid	Formula	pH of acid with concentration of 1 mol/dm³	Acid	Formula	pH of acid with concentration of 1 mol/dm³
A		2.42	**C**		0.65
B		1.44	**D**		0.32

05.1 Deduce the effect of the presence of chlorine atoms on the degree of ionisation of a carboxylic acid. Justify your answer. **[3 marks]**

05.2 A student adds drops of acids **A** and **D**, separately, onto lumps of calcium carbonate. Compare the predicted observations for the two acids. **[2 marks]**

Exam Tip

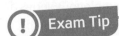

pH is a value showing how many H⁺ ions have dissociated from the compound. The more H⁺ ions that dissociate the more ionised the carboxylic aid will be.

05.3 When ethanoic acid reacts with calcium carbonate, the salt made is calcium ethanoate. Predict the name of the salt made when the acid with the formula CH_3CH_2COOH reacts with calcium carbonate.
[1 mark]

06 A student carried out some reactions of three organic compounds pentene, ethanoic acid, and ethanol. The compounds were labelled **X**, **Y**, and **Z**. The student did not know which compound was which. Their observations are shown in **Table 3**.

Table 3

Compound	Add sodium	Add sodium carbonate	Burn the compound
X	bubbles slowly	no reaction	burns with a blue flame that is hard to see
Y	no reaction	no reaction	burns with a smoky flame
Z	bubbles quickly	bubbles	

06.1 Deduce the identities of **X**, **Y**, and **Z**. Justify your decisions. **[6 marks]**

06.2 The student carries out another experiment. They boil some propanol with an oxidising agent. Name the homologous series that propanol is in. **[1 mark]**

06.3 Deduce the name of the organic product of the reaction of propanol with the oxidising agent. **[1 mark]**

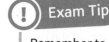

Exam Tip

Remember to use the evidence from **Table 3** to support your reasoning.

07 A student wanted to compare the energy transferred when different alcohols burn in air. They measured the temperature change of the water when different alcohols were burnt in a spirit burner. **Figure 2** shows the apparatus.

07.1 Identify the independent variable in the investigation. **[1 mark]**

07.2 Suggest **two** control variables in the investigation. **[2 marks]**

07.3 **Table 4** shows the student's results. Explain the trend shown by the results. **[1 mark]**

07.4 Write a balanced chemical equation for the complete combustion of butanol, C_4H_9OH. **[3 marks]**

07.5 Pentanol has five carbon atoms. Complete the structure of pentanol. **[1 mark]**

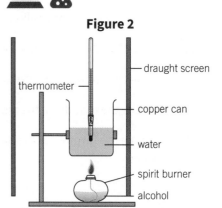

Figure 2

draught screen

thermometer

copper can

water

spirit burner

alcohol

Table 4

Alcohol	Temperature increase of the water in °C
methanol	6.5
ethanol	9.5
propanol	13.0
butanol	16.5

Exam Tip

There are always the same two products from complete combustion. What will change is how the equation is balanced. Start with the carbons, then the hydrogens, leaving the oxygens to last.

08 Ethanol can be made in a fermentation reaction. The equation for the reaction is:

$$C_6H_{12}O_6 \rightarrow 2C_2H_5OH + 2CO_2$$

! Exam Tip

Carbon will always make four bonds, whilst hydrogens will always make one bond and oxygens will make two bonds.

08.1 Name the living organism required for the fermentation reaction to occur. **[1 mark]**

08.2 Give the temperature at which fermentation is normally carried out. **[1 mark]**

08.3 Give **two** uses of ethanol. **[2 marks]**

08.4 In a fermentation reaction, 80.0 g of ethanol is made. Calculate the mass of glucose that reacted. **[5 marks]**

! Exam Tip

The first step is to determine the M_r of glucose and ethanol.

09 Three compounds have the formulae **X**: $C_{10}H_{20}$, **Y**: $C_{10}H_{22}$, **Z**: $C_{10}H_{21}OH$

09.1 Calculate the relative formula mass of compound **Z**.
Relative atomic masses A_r: C = 12; H = 1; O = 16 **[2 marks]**

! Exam Tip

Use the general formulae to determine which homologous series the compounds belong to.

09.2 Give the letter of the compound that reacts with steam to make an alcohol. **[1 mark]**

09.3 Write the formula of the product of the reaction of compound **X** with hydrogen. **[1 mark]**

09.4 Name the type of compound formed when compound **Z** reacts with ethanoic acid. **[1 mark]**

10 This question is about alkanes and alkenes.

10.1 Name the alkane with the formula C_4H_{10}. **[1 mark]**

10.2 Name the compound with the formula shown in **Figure 3**. **[1 mark]**

10.3 Deduce the formula of the alkene that has 8 carbon atoms. **[1 mark]**

10.4 Deduce the formula of the alkane that has 12 hydrogen atoms. **[1 mark]**

Figure 3

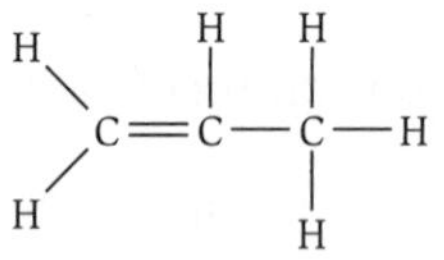

11 **Table 5** shows the solubility of some alcohols in water.

Table 5

Number of carbon atoms in alcohol	Name of alcohol	Solubility in g/per 100 g of water
1		the alcohol completely mixes with water
2	ethanol	
3	propanol	
4	butanol	8.14
5	pentanol	2.64
6	hexanol	0.592
7	heptanol	0.0928

11.1 Name the alcohol with one carbon atom. **[1 mark]**

11.2 Describe the pattern shown in **Table 5**. **[2 marks]**

11.3 Suggest a reason for the pattern shown in **Table 5**. **[1 mark]**

11.4 Calculate the solubility of pentanol in mol/dm³. Give your answer to two significant figures. Relative atomic masses A_r: C = 12; H = 1; O = 16 **[5 marks]**

12 The reaction between hydrochloric acid and sodium hydroxide is exothermic.

12.1 What type of reaction is this? Choose **one** answer. **[1 mark]**

decomposition   electrolysis   neutralisation   reduction

12.2 A student makes the following conclusion about the reaction.

"As the reaction happened, the test tube got hotter. This shows that more energy was produced."

Is the student correct? Explain your answer. **[3 marks]**

12.3 Describe the energy transfers involved in the reaction between hydrochloric acid and sodium hydroxide in terms of bonds, and how this makes the reaction exothermic. **[3 marks]**

12.4 Give **one** other example of an exothermic reaction. **[1 mark]**

13 A student investigated the thermal decomposition of calcium carbonate, $CaCO_3$. They heated 40 g of calcium carbonate in a test tube.

13.1 Calculate the relative formula mass of calcium carbonate. Relative atomic masses A_r: C = 12; H = 1; O = 16; Ca = 40 **[2 marks]**

13.2 Calculate the number of moles of calcium carbonate that the student heated. **[2 marks]**

13.3 Complete the balanced symbol equation for the reaction. **[1 mark]**
$$CaCO_3(s) \rightarrow \underline{\hspace{1cm}}(s) + CO_2(g)$$

13.4 Calculate the percentage atom economy for the reaction to produce calcium oxide. Give your answer to two significant figures. **[4 marks]**

> ! Exam Tip
>
> Take each of the elements in carbon dioxide away from calcium carbonate and see what you have left over

13.5 The student measures the mass at the end of the reaction. The mass has decreased. Explain why the mass has decreased. **[2 marks]**

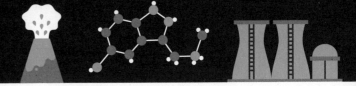

C16 Polymers

Polymers

Polymers are very long molecules made up of lots of smaller molecules joined together in a repeating pattern. The smaller molecules are called **monomers**. The process of turning many monomers into a polymer is called polymerisation.

There are two main types of polymerisation.

Type of polymerisation	Monomers	Products of polymerisation
addition polymerisation	molecules with C=C bonds, such as alkenes	just the polymer
condensation polymerisation	diols, dicarboxylic acids, or diamines	polymer and water

Addition polymerisation

Addition polymerisation starts with molecules with a C=C bond (e.g., alkenes) as the monomer. The carbon-carbon double bond breaks in each molecule, and the carbon atoms then link together.

$$n \quad \underset{\substack{\text{many single} \\ \text{ethene monomers}}}{\overset{\begin{array}{cc}H & H \\ | & | \\ C = C \\ | & | \\ H & H\end{array}}{}} \longrightarrow \underset{\substack{\text{long chain} \\ \text{of poly(ethene)}}}{\overset{\begin{array}{cc}H & H \\ | & | \\ \left(C - C \right)_n \\ | & | \\ H & H\end{array}}{}} \quad \text{where } n \text{ is a large number}$$

The n refers to a large number of molecules. The rounded brackets and the bonds sticking out of them represent where the next molecule in the chain goes.

The inside of the brackets is known as the **repeating unit** – the section that repeats over and over again many thousands of times in the polymer.

Addition polymers are named after the monomer used to create them.

- An addition polymer made of ethene is called poly(ethene).
- An addition polymer made of propene is called poly(propene).

Natural polymers

Amino acids and proteins

Condensation reactions can also happen with just one monomer molecule, so long as the molecule has two different functional groups.

Amino acids have an **amine** functional group and a carboxylic acid functional group. The amine functional group has a nitrogen bonded to a carbon and two hydrogens.

Glycine is the simplest amino acid.

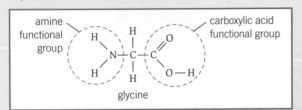
glycine

When many molecules of glycine react together they form a **polypeptide**.

There are many different types of amino acids. They can react together to form many different polypeptides. When lots of polypeptides come together they form something called a **protein**.

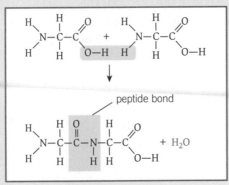

DNA

All genetic information is stored in **DNA**. Genetic information contains the instructions for the functioning and development of living organisms.

DNA is made of two long polymers that wind around each other in a double helix. The polymers are made of four different monomers called **nucleotides**.

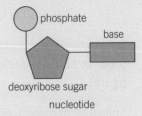
nucleotide

Starch and cellulose

Starch and cellulose are another two **natural polymers**. Both of these are made from glucose molecules joined together. Whether the resulting polymer is starch or cellulose depends on how the glucose molecules form chains with each other.

Condensation polymerisation

Condensation polymerisation can involve two different monomers, each has *two* functional groups.

Molecule **A** is a diol. It has two –OH groups: one at either end.

Molecule **B** is a **dicarboxylic acid**. It has a carboxylic acid group at either end.

To simplify the diagrams, a rectangle is used to represent C—C.

Molecule **A**

Molecule **B**

When molecule **A** and molecule **B** react together, the –OH group from the carboxylic acid and a hydrogen atom from the –OH group on the alcohol join together to form water.

+ H_2O

Another molecule **B** and another molecule **A** can now react with either side of the molecule that has been formed.

+ $2H_2O$

You could keep adding more molecules in the pattern ABABABABA. Every time a molecule is added, a water molecule is produced. This type of reaction is called a **condensation reaction**.

If you keep adding molecules, a condensation polymer is produced. This is represented by:

$$n \text{ HO}-\boxed{}-\text{OH} + n \text{ HOOC}-\boxed{}-\text{COOH} \longrightarrow \left(\text{O}-\boxed{}-\text{O}-\text{CO}-\boxed{}-\text{CO}\right)_n + 2n\text{H}_2\text{O}$$

When diols (compounds with two –OH groups) and dicarboxylic acids react together, they form polyesters.

🔑 **Key terms**

Make sure you can write a definition for these key terms.

addition polymerisation	amine	amino acid	condensation polymerisation	DNA
monomer	natural polymer	nucleotide	polymer	polypeptide	protein	repeating unit

Retrieval

Learn the answers to the questions below then cover the answers column with
a piece of paper and write as many as you can. Check and repeat.

	C16 questions	Answers
1	What are monomers?	small molecules that join together to form a long chain
2	What is a polymer?	a very long molecule made of repeating units
3	What is a repeating unit?	the smallest part of a polymer that repeats itself throughout the chain
4	What is polymerisation?	a reaction that turns multiple monomers into polymers
5	What are the two types of polymerisation?	addition and condensation
6	What kind of monomers are involved in addition polymerisation?	molecules with $C=C$ bonds, such as alkenes
7	What kind of monomers are involved in condensation polymerisation?	monomers with two functional groups
8	What other products are made in condensation polymerisation?	water (normally)
9	What does n represent in an equation showing polymerisation?	a very large number
10	What is a natural polymer?	a polymer that is produced naturally by organisms
11	Give four examples of natural polymers.	polypeptides, starch, cellulose, DNA
12	What are amino acids?	the building blocks for polypeptides and proteins, which have an amine and a carboxylic acid group
13	What is a polypeptide?	a polymer made from many amino acids
14	What is a protein?	a polymer made from amino acids
15	Which monomer makes up starch and cellulose?	glucose
16	What is DNA?	a molecule containing genetic information
17	Which monomers are DNA made of?	nucleotides
18	How is DNA arranged?	double helix

Put paper here

Now go back and use the questions below to check your knowledge from previous chapters.

C16

Previous questions

Answers

1	What is dynamic equilibrium?		the point in a reversible reaction when the rate of the forward and reverse reactions are the same
2	What is a homologous series?	Put paper here	a group of compounds with the same functional group
3	When an acid reacts with a metal, why does the mass decrease?		a gas is produced and escapes
4	How can concentration in mol/dm³ be calculated?	Put paper here	$\dfrac{\text{moles of solute}}{\text{volume (dm}^3\text{)}}$
5	How is percentage yield calculated?		$\left(\dfrac{\text{actual yield}}{\text{theoretical yield}}\right) \times 100$

 # Maths Skills

Practise your maths skills using the worked example and practice questions below.

Gradients	Worked Example	Practice
The gradient of a straight line on a graph tells you how steep it is. To determine the gradient: **1** pick two points on the line **2** subtract the smallest y-axis value from the largest y-axis value of your two points **3** take the two x-axis values from your two points and subtract the smallest from the largest **4** divide your answer from **2** by your answer from **3**. If you have a positive correlation, your gradient will be positive. If you have a negative correlation, your gradient will be negative.	The graph below shows how the mass of solid copper objects varies with their volume. Calculate the gradient of the graph. 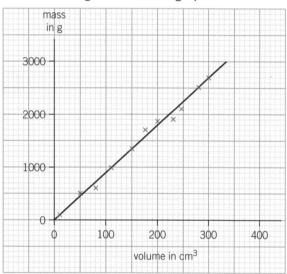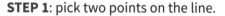 **STEP 1**: pick two points on the line. **STEP 2**: subtract the y-axis values: $2500 - 1000 = 1500$ **STEP 3**: subtract the x-axis values: **STEP 4**: divide the y value by the x value: $\dfrac{1500}{170} = 8.8$ g/cm³	The graph below shows the temperature over time of a liquid being heated. Calculate the gradient of the graph. 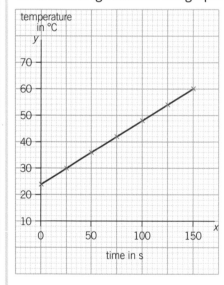

Exam-style questions

01.1 Draw **one** line from the name of each polymer (or type of polymer) to the type of polymerisation used to make the polymer. **[4 marks]**

Polymer or type of polymer	Type of polymerisation used to make the polymer
poly(ethene)	
polyester	addition
poly(propene)	condensation
polypeptide	

> **! Exam Tip**
>
> There will be more than one line going into each box on the right-hand side.

01.2 **Figure 1** shows the repeating unit of a polymer.

Figure 1

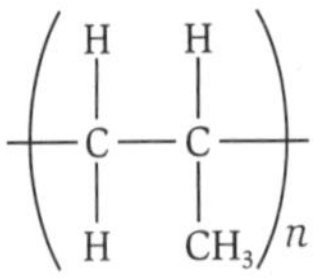

$$\left(\begin{array}{cc} \overset{\displaystyle H}{\underset{\displaystyle H}{|}} & \overset{\displaystyle H}{\underset{\displaystyle CH_3}{|}} \\ -C - C- \end{array}\right)_n$$

Which monomer is needed to make the polymer in **Figure 1**?
Tick **one** box. **[1 mark]**

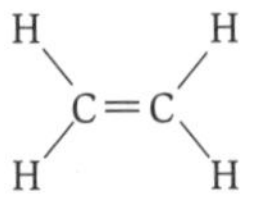

 ☐

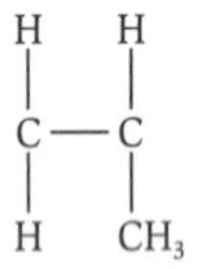

 ☐

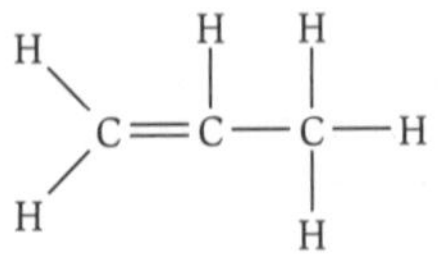

 ☐

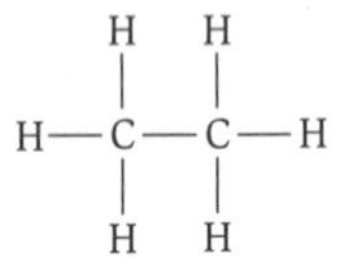

 ☐

> **! Exam Tip**
>
> On **Figure 1** draw another line between the two carbons, then scribble out the lines touching the brackets – now try to match it to one of the answer options.

01.3   Give the name of the polymer formed from the repeating unit shown in **Figure 2**.   [1 mark]

Figure 2

Exam Tip

First identify the monomer. This will help you to name the polymer.

_____

02   **Table 1** shows the monomers used to make four polymers: **W**, **X**, **Y**, and **Z**.

Table 1

Polymer	Monomer(s)
W	glucose
X	butene
Y	propane diol butanedioic acid
Z	glycine lysine tyrosine

02.1   Give the letter of the polymer that is a polypeptide.   [1 mark]

_____

02.2   Give the letter of **one** polymer that could be starch or cellulose.   [1 mark]

_____

Exam Tip

Try to write the names of each of the polymers that will be formed from these monomers – this will help you with the rest of this question.

02.3   Give the letters of **two** naturally occurring polymers that are formed in condensation reactions.   [1 mark]

_____

02.4   Name the type of polymerisation that forms polymer **X**.   [1 mark]

_____

02.5   Give the name of polymer **X**.   [1 mark]

_____

03   This question is about condensation polymerisation.

03.1   Give the number of functional groups that a monomer must have in order to take part in a condensation polymerisation reaction.

[1 mark]

03.2   Condensation polymerisation reactions result in two products, the polymer and a small molecule. Name **one** small molecule that is often made in condensation polymerisation reactions.   [1 mark]

03.3   Which pair of monomers react together to make the polymer with the repeating unit shown in **Figure 3**?

Figure 3

$(-\square-OOC-\square-COO-)_n$

Choose **one** answer.   [1 mark]

HO—$\square$—OH and HO—$\square$—OH

HO—$\square$—OH and HOOC—$\square$—COOH

HOOC—$\square$—COOH and HOOC—$\square$—COOH

HO—$\square$—OH and H_2O

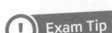

Exam Tip

Think about what you normally see formed in condensation; the small molecule formed in condensation polymerisation is often the same.

Exam Tip

In **Figure 3** draw a line where you think the link has been made and use that to try and work out what the monomers are.

04   **Figure 4** shows a section of a polymer. The polymer was made by addition polymerisation.

Figure 4

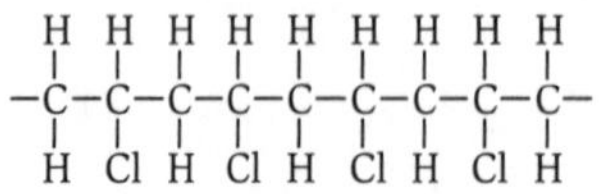

$$-\overset{\overset{H}{|}}{C}-\overset{\overset{H}{|}}{\underset{\underset{Cl}{|}}{C}}-\overset{\overset{H}{|}}{\underset{\underset{H}{|}}{C}}-\overset{\overset{H}{|}}{\underset{\underset{Cl}{|}}{C}}-\overset{\overset{H}{|}}{\underset{\underset{H}{|}}{C}}-\overset{\overset{H}{|}}{\underset{\underset{Cl}{|}}{C}}-\overset{\overset{H}{|}}{\underset{\underset{H}{|}}{C}}-\overset{\overset{H}{|}}{\underset{\underset{Cl}{|}}{C}}-\overset{\overset{H}{|}}{\underset{\underset{H}{|}}{C}}-$$

04.1   Draw the repeating unit of the polymer.   [1 mark]

04.2   Draw the formula of the monomer used to make the polymer.

[1 mark]

Exam Tip

Draw circles on **Figure 4** to identify the repeating unit, then redraw that with a double bond in the middle.

04.3   Suggest the percentage atom economy of the formation of the polymer from its monomer. Justify your answer.   **[2 marks]**

05   Nylon is a condensation polymer. It is made from the two monomers shown in **Figure 5**.

Figure 5

05.1   When one repeating unit of nylon is made from the monomers in **Figure 5**, two molecules of HCl are also made. Draw the repeating unit of nylon.   **[2 marks]**

05.2 **Figure 6** represents the formula of an amino acid. It polymerises to make a polypeptide. When one repeating unit is made, one molecule of H_2O is also made.

Figure 6

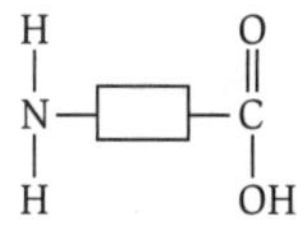

Draw the repeating unit of the polymer made from the monomer in **Figure 6**. **[2 marks]**

Exam Tip

Identify which parts of the compound shown in **Figure 6** will form the small H_2O molecule that is not part of the repeating unit.

05.3 Identify **one** difference between the monomers in **Figure 5** and the monomer in **Figure 6**. **[2 marks]**

06 PTFE is a polymer made form the monomer shown in **Figure 7**.

Figure 7

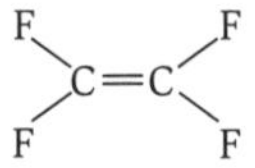

06.1 Identify the feature in **Figure 7** that means it can undergo addition polymerisation. **[1 mark]**

06.2 Draw a section of the polymer made from four monomer molecules. **[1 mark]**

06.3 The name of the monomer is tetrafluoroethene. Deduce the name of the polymer it forms. **[1 mark]**

Exam Tip

Don't be put off by the fluorines instead of hydrogens; it acts in exactly the same way. You just need to apply what you know to a new context.

07 This question is about DNA.

07.1 Describe the function of DNA. **[1 mark]**

07.2 Give the number of nucleotide monomers that form the two polymer chains in DNA. **[1 mark]**

07.3 One type of nucleotide monomer includes a base called guanine. The formula of guanine is $C_5H_5N_5O$. Calculate the relative formula mass of guanine. Relative atomic masses A_r: C = 12, H = 1, N = 14, O = 16 **[2 marks]**

08 **Figure 8** shows part of a polymerisation reaction. Some of the formulae are incomplete, and one product is not given.

Figure 8

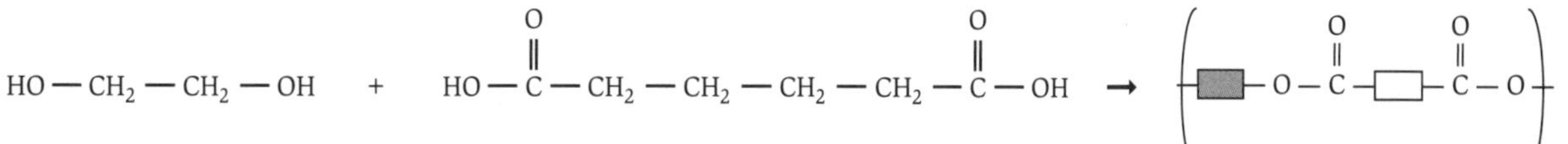

monomer **A**   monomer **B**

08.1 Name the functional groups present in monomers **A** and **B**.
[2 marks]

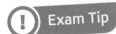
Exam Tip

Circle the functional groups to help you pick them out from the rest; use a different colour per functional group.

08.2 Name the type of polymer that is formed in the reaction. [1 mark]

08.3 Give the formula of the other product formed in the reaction.
[1 mark]

08.4 Deduce the number of carbon atoms that are joined together in the shaded box in **Figure 8**. [1 mark]

Exam Tip

List the numbers of carbons, hydrogens, and oxygens first so you don't miss any out.

08.5 Calculate the relative formula mass of the repeating unit of the polymer. Relative atomic masses A_r: C = 12, H = 1, O = 16
[2 marks]

09 This question is about addition and condensation polymerisation reactions.

09.1 Name **one** polymer that is made in an addition polymerisation reaction. [1 mark]

Exam Tip

The question has given you the structure of the answer so make sure you cover every part.

Condensation polymerisation: the monomers required

Condensation polymerisation: products formed

Addition polymerisation: the monomers required

Addition polymerisation: products formed

09.2 Name **one** type of naturally occurring compound that is made from a condensation polymerisation reaction. [1 mark]

09.3 Compare addition and condensation polymerisation. In your answer, include information about the monomers required and products formed in each type of polymerisation. [6 marks]

10 Some students carried out a titration between sodium hydroxide and sulfuric acid. They used the following method:

1 Use a measuring cylinder to transfer 25.0 cm³ of sodium hydroxide to a conical flask.

2 Fill a burette with acid.

3 Add 1 cm³ of indicator to the flask.

4 Add acid from the burette to the flask until the indicator changes colour.

5 Repeat the procedure, adding the acid drop by drop as the end point approaches.

10.1 Suggest an improvement to step **1**. [1 mark]

Exam Tip

Think of a more precise piece of equipment you could use.

10.2 Suggest **two** things the students can do to avoid spillages in step **2**.
[2 marks]

10.3 Identify the mistake in step **3**. [1 mark]

10.4 **Table 2** shows their results.

Table 2

Titration	Initial burette reading in cm³	Final burette reading in cm³	Volume of acid added in cm³
1	1.30	18.35	19.65
2	19.65	37.85	18.20
3	2.40	20.70	18.30
4	20.70	39.95	
5	0.05	18.30	18.25

Write down the missing volume in **Table 2**.   **[1 mark]**

10.5 Use the students' results to calculate the mean volume of acid added.   **[1 mark]**

10.6 The equation for the reaction is:

$$H_2SO_4(aq) + 2NaOH(aq) \rightarrow Na_2SO_4(aq) + 2H_2O(l)$$

The concentration of acid used was 0.100 mol/dm³. Calculate the concentration of sodium hydroxide in mol/dm³. Give your answer to three significant figures.   **[5 marks]**

11 A chemist measured the diameters of some particles. **Table 3** shows the results.

Table 3

Particle	Diameter in nm
W	25
X	3400
Y	60
Z	950

11.1 Identify the coarse particle in **Table 3**.   **[1 mark]**

11.2 Identify the particle in **Table 3** that has the greatest surface-area-to-volume ratio.   **[1 mark]**

11.3 Write the diameter of particle **X** in standard form. Give your answer in nanometres.   **[1 mark]**

11.4 Determine the diameter of particle **Z** in metres. Give your answer in standard form.   **[2 marks]**

12 Heptene, heptanol, and heptane are three compounds with 7 carbon atoms.

12.1 Each substance is mixed with bromine water. Which substance will turn the bromine water colourless?   **[1 mark]**

12.2 Which compound will form heptanoic acid when boiled with potassium dichromate(VI) solution?   **[1 mark]**

12.3 Write the chemical formula of heptane.   **[1 mark]**

12.4 Compare boiling point of heptane with ethane. Explain your answer.   **[4 marks]**

 **Exam Tip**

Only use the concordant results.

 Exam Tip

Highlight all the important bits you need for this and keep them in one place; this will save you needing to read over the text every time you need a number.

Volume of acid = mean titre

Volume of alkali = 25.0cm³

Concentration of acid = 0.100 mol/dm³

Ratio of acid:alkali = 1:2

 **Exam Tip**

Look at the differences in the ending of the names.

 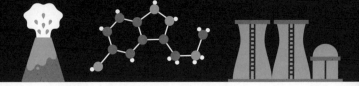

C17 Chemical analysis

Pure and impure

In chemistry, a **pure** substance contains a single element or compound that is not mixed with any other substance.

Pure substances melt and boil at specific temperatures.

An **impure** substance contains more than one type of element of compound in a **mixture**.

Impure substances melt and boil at a range of temperatures.

Formulations

Formulations are examples of mixtures. They have many different components (substances that make them up) in very specific proportions (amounts compared to each other).

Scientists spend a lot of time trying to get the right components in the right proportions to make the most useful product.

Formulations include fuels, cleaning agents, paints, alloys, fertilisers, and foods.

Chromatography

Chromatography is a method to separate different components in a mixture. It is set up as shown here, with a piece of paper in a beaker containing a small amount of solvent.

The **R_f value** is a ratio of how far up the paper a certain spot moves compared to how far the **solvent** has travelled.

$$R_f = \frac{\text{distance moved by substance}}{\text{distance moved by solvent}}$$

It will always be a number between 0 and 1.

The R_f value depends on the solvent and the temperature, and different substances will have different R_f values. The R_f values for particular solvents can be used to identify a substance.

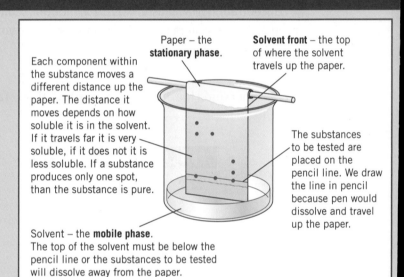

Paper – the **stationary phase**.

Solvent front – the top of where the solvent travels up the paper.

Each component within the substance moves a different distance up the paper. The distance it moves depends on how soluble it is in the solvent. If it travels far it is very soluble, if it does not it is less soluble. If a substance produces only one spot, than the substance is pure.

The substances to be tested are placed on the pencil line. We draw the line in pencil because pen would dissolve and travel up the paper.

Solvent – the **mobile phase**. The top of the solvent must be below the pencil line or the substances to be tested will dissolve away from the paper.

Testing gases

Common gases can be identified using the follow tests:

Gas	What you do	What you observe if gas is present
hydrogen	hold a lighted splint near the gas	hear a squeaky pop
oxygen	hold a glowing splint near the gas	splint re-lights
carbon dioxide	bubble the gas through limewater	the limewater turns milky (cloudy white)
chlorine	hold a piece of damp litmus near the gas	bleaches the litmus white

 **Key terms**

Make sure you can write a definition for these key terms.

chromatography   flame emission spectroscopy   flame test   formulation

impure   instrumental analysis   mobile phase   precipitate   pure

R_f value   solvent   solvent front   stationary phase

Testing for cations

Metal ions always have a positive charge (i.e., they are cations). Sodium hydroxide solution can be used to identify some metal ions.

Cation	Positive result
aluminium ions, Al^{3+}	on slow addition of excess sodium hydroxide solution, white **precipitate** forms that eventually dissolves again with excess sodium hydroxide
calcium ions, Ca^{2+}	on addition of excess sodium hydroxide solution, white precipitate that does not dissolve
magnesium ions, Mg^{2+}	on addition of excess sodium hydroxide solution, white precipitate that does not dissolve
copper(II) ions, Cu^{2+}	forms a blue precipitate
iron(II) ions, Fe^{2+}	forms a green precipitate
iron(III) ions, Fe^{3+}	forms a brown precipitate

Testing for anions

Anion	Test	Positive result
carbonate, CO_3^{2-}	add dilute acid	carbon dioxide gas formed which can be test for with limewater
chloride, Cl^-	add silver nitrate solution in the presence of nitric acid	white precipitate formed
bromide, Br^-	add silver nitrate solution in the presence of nitric acid	cream precipitate formed
iodide, I^-	add silver nitrate solution in the presence of nitric acid	yellow precipitate formed
sulfate, SO_4^{2-}	add barium chloride solution in the presence of hydrochloric acid	white precipitate formed

Flame tests

Substances containing metals can produce a coloured light in a flame. This can be used to identify the metal. However, if there is more than one metal in the substance then this method will not work, as the colours mix and intense colours mask more subtle colours.

Metal	Flame colour
lithium	crimson
sodium	yellow
potassium	lilac
calcium	orange-red
copper	green

Instrumental methods

Instrumental analysis involves using complex scientific equipment to test substances.

Instrumental methods are rapid and accurate. They are also sensitive, which means they can give results even with very small amounts of substance.

Flame emission spectroscopy

Flame emission spectroscopy is a type of instrumental analysis similar to a **flame test**.

The sample solution is put into a flame and the light given off is passed through a spectroscope. Instead of a human observing a colour, the instrument tells you exactly which wavelength of light is being given off as a line spectrum. You can then compare the spectrum to a reference to establish the identity of your sample. You can also measure the concentration of the substance in your sample solution.

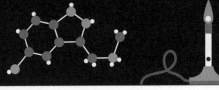

Learn the answers to the questions below then cover the answers column with
a piece of paper and write down as many as you can. Check and repeat.

	C17 questions	Answers
1	In chemistry, what is a pure substance?	something made of only one type of substance
2	What is the difference between the melting and boiling points of a pure and impure substance?	pure – sharp/one specific temperature impure – broad/occurs across a range of temperatures
3	What is a formulation?	a mixture designed for a specific purpose
4	What are some examples of formulations?	fuels, cleaning agents, paints, medicines, alloys, fertilisers, and foods
5	What is chromatography?	a process for separating coloured mixtures
6	How is R_f calculated?	$R_f = \dfrac{\text{distance moved by substance}}{\text{distance moved by solvent}}$
7	What is the test for hydrogen?	a lit splint gives squeaky pop
8	What is the test for oxygen?	re-lights a glowing splint
9	What is the test for carbon dioxide?	turns limewater milky if bubbled through it
10	What is the test for chlorine?	bleaches damp litmus paper
11	What is the test for aluminium, calcium, and magnesium ions?	forms white precipitate with sodium hydroxide solution
12	How can aluminium ions be distinguished from calcium and magnesium ones?	the white precipitate will dissolve with excess sodium hydroxide
13	What colour precipitates are formed when sodium hydroxide solution is added to solutions of copper(II), iron(II), and iron(III) ions?	copper(II) ions form blue precipitate, iron(II) ions form green precipitate, iron(III) ions form brown precipitate
14	What is the test for a halide ion?	add silver nitrate and nitric acid: chloride forms white precipitate, bromide forms cream precipitate, iodide forms yellow precipitate
15	What is the test for a carbonate ion?	carbon dioxide gas formed on addition of acid
16	What is the test for a sulfate ion?	white precipitate formed with hydrochloric acid and barium chloride
17	What colours are produced by different metals in a flame test?	lithium – crimson; sodium – yellow; potassium – lilac; calcium – orange-red; copper – green
18	What is instrumental analysis?	using complex scientific equipment to identify substances
19	What are the three advantages of instrumental analysis?	rapid, accurate, and sensitive
20	What information does flame emission spectroscopy produce?	the wavelength of light given off by a metal in a flame to identity of the metal and its concentration

(Put paper here — printed vertically down the centre column)

Now go back and use the questions below to check your knowledge from previous chapters.

C17

Previous questions | Answers

	Previous questions		Answers
1	What is polymerisation?		a reaction that turns multiple monomers into polymers
2	In terms of bond breaking and making, what is an exothermic reaction?	Put paper here	less energy is required to break the bonds than is released when making the bonds
3	What effect does increasing pressure have on the rate of reaction?		Increases
4	What is electrolysis?	Put paper here	process of using electricity to extract elements from a compound
5	In terms of pH, what is a neutral solution?		a solution with a pH of 7
6	What is a neutralisation reaction?	Put paper here	a reaction between an acid and a base to produce water
7	How is the amount of H^+ ions in a solution related to its pH?		the more H^+ ions, the lower the pH
8	In terms of electrons, what is oxidation?		loss of electrons

Required Practical Skills

Practise answering questions on the required practicals using the example below.
You need to be able to apply your skills and knowledge to other practicals too.

Identifying ions	Worked Example	Practice
You need to be able to describe how to identify unknown compounds, including: • flame tests for different ions • metal ion precipitation test • carbonate test • sulfate test • halide test A common exam question asks you to identify mystery compounds based on their results in the above tests, so you need to know their methods and the observations for positive results. You also need to be able to write the formula for the ions these tests identify, and write equations using them.	A student wanted to test a sample for sulfate and halide ions. Write a method that the student could use to test for these two ions. **Answer**: Split the sample between two test tubes. In one test tube, carry out the test for sulfate ions by adding a few drops of dilute hydrochloric acid and dilute barium chloride solution. If a white precipitate appears, the sample contains sulfate ions. In the other test tube, carry out the test for halide ions by adding a few drops of dilute nitric acid followed by dilute silver nitrate solution. Formation of a white, yellow, or cream precipitate indicates the presence of halide ions.	1 When testing for halides it is a good idea to test three known samples to use as a reference. Explain why. 2 A sample produced a lilac colour in a flame test, and effervesced when treated with hydrochloric acid. Identify the formula of the sample. 3 Describe why the colour of the flame test for sodium is hard to see, and suggest a way to resolve this problem.

Exam-style questions

01 A student pours dilute hydrochloric acid into a test tube and adds magnesium ribbon. A gas is formed that the student collects.

01.1 Name the **two** products formed in the reaction. **[2 marks]**

1 _____

2 _____

01.2 Describe the test the student can carry out to identify the gas collected. Give the expected result. **[2 marks]**

Test: _____

Expected result: _____

01.3 Another student reacted magnesium with sodium carbonate. A gas was formed that the student collected.

Describe the test the student can carry out to identify the gas collected. Give the expected result. **[2 marks]**

Test: _____

Expected result: _____

01.4 The scientist had a sample of a pale green gas. They inserted a glowing splint into the gas. The splint went out.

The scientist then put some damp litmus paper into the gas. The litmus paper turned white.

Identify the gas. **[1 mark]**

_____

02 **Table 1** shows the melting points for four substances purchased from a grocery store.

Table 1

Substance	Melting point in °C
A	90–95
B	0
C	30–32
D	47

All four substances are described as 'pure' on their containers.

02.1 Describe the difference between a chemically pure substance and a substance described as pure in everyday language. **[2 marks]**

_____

_____

_____

_____

02.2 Identify the **two** substances that are chemically pure in **Table 1**.

[1 mark]

1 _____

2 _____

02.3 A washing detergent was also purchased. The laundry detergent contains:

- a chemical that removes grease
- a colouring
- a fragrance.

Explain why the laundry detergent is a formulation. [2 marks]

_____

_____

_____

03 A student has three metal compounds. They mix a sample of each compound with sodium hydroxide.

03.1 Metal compound **A** produces a white precipitate. When more sodium hydroxide solution is added, the precipitate dissolves. Identify the metal. [1 mark]

03.2 Metal compound **B** produces a blue precipitate. Identify the metal ion. [1 mark]

03.3 Metal compound **C** produces a white precipitate. It does not dissolve on addition of further sodium hydroxide solution. Identify the metal. [1 mark]

03.4 The student then carries out some tests to identify the non-metal ion in each compound. **Table 2** shows their results.

> **! Exam Tip**
>
> You need to recall what happens when a small amount of sodium hydroxide is added, and what happens when it is in excess. There are two results needed for a reaction with sodium hydroxide.

> **! Exam Tip**
>
> It will be the same non-metal ions for all three.

Table 2

Metal compound	Addition of barium chloride	Addition of dilute hydrochloric acid
A	white precipitate	no change
B	white precipitate	no change
C	white precipitate	no change

Identify the non-metal ion in all three compounds. [1 mark]

03.5 Write a balanced symbol equation for the reaction in **03.2**.

[3 marks]

04 All plants carry out photosynthesis. A student sets up the apparatus shown in **Figure 1** to investigate the products of photosynthesis.

Figure 1

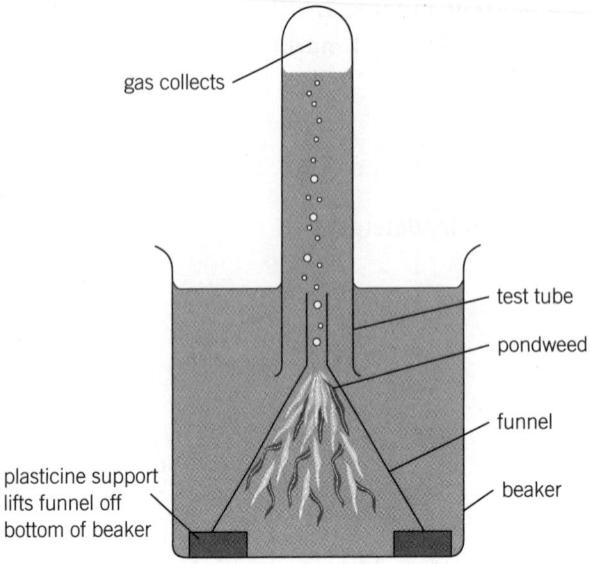

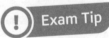

The student collects some of the gas in the test tube and inserts a glowing split. The splint relights.

04.1 Use the student's result to complete and balance the symbol equation for photosynthesis. **[2 marks]**

_____ CO_2 + _____ H_2O → $C_6H_{12}O_6$ + _____

Exam Tip

Only fill in the gaps. If you're tempted to try to write an answer outside the gaps, STOP!

04.2 Photosynthesis occurs in the leaves of plants. Pigments in the leaves help the process to occur.

A student uses paper chromatography to investigate the pigments in a leaf.

Describe a method to carry out the paper chromatography experiment. In your answer name any equipment required. **[6 marks]**

04.3 The student finds that there are four pigments in the leaves. Sketch the chromatogram that the student has produced. **[1 mark]**

04.4 The student wants to calculate the R_f value of one of the pigments in the chromatogram. Give the equation to calculate the R_f value. **[1 mark]**

04.5 The solvent travelled 12.0 cm. One of the spots travelled 8.6 cm. Use **Table 3** to identify which pigment was responsible for the spot.

Table 3

Pigment	R_f value
carotene	0.95
xanthophyll	0.72
chlorophyll a	0.65
chlorophyll b	0.45

Exam Tip

R_f value =
$$\frac{\text{distance moved by spot}}{\text{distance moved by solvent}}$$

05 Chemists use different chemical tests to identify the substances in a compound.

05.1 A chemist had three unknown gases. The chemist carried out three simple tests to identify the gases. Their observations are shown in **Table 4**.

Table 4

Gas	Test		
	Burning splint held at open end of tube	Glowing splint inserted into tube	Bubbled through limewater
A	no observation	no observation	cloudy
B	no observation	splint relights	no change
C	pop sound	no observation	no change

Identify the gases **A**, **B**, and **C**. **[3 marks]**

05.2 The chemist has a fourth gas. The chemist thinks the gas is chlorine. Describe how the chemist could confirm that the gas is chlorine. **[2 marks]**

05.3 The chemist also has a sample of lithium bromide, magnesium bromide, and lithium carbonate. All three compounds are white solids. The compounds are not labelled.

Describe an experimental procedure the chemist could use to identify each compound. Your procedure should use as few tests as possible. Include the expected results for each test. **[6 marks]**

Exam Tip

This is 3 marks for 3 short answers, don't be tempted to explain you reasoning because you won't get any extra marks and you'll just be wasting time.

06 A student investigated the dyes in three felt tip pens. The dyes are soluble in water. They set up a chromatography experiment. **Figure 2** shows the apparatus the students used.

Figure 2

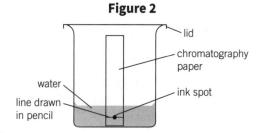

06.1 Name the mobile phase in this chromatography experiment. **[1 mark]**

06.2 The student made a mistake in setting up the apparatus. Identify the mistake and give **one** problem caused by this mistake. **[2 marks]**

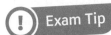

Exam Tip

Look carefully at the diagram to find the mistake.

06.3 Another student set up the apparatus correctly. **Figure 3** shows the chromatogram the student obtained.

Figure 3

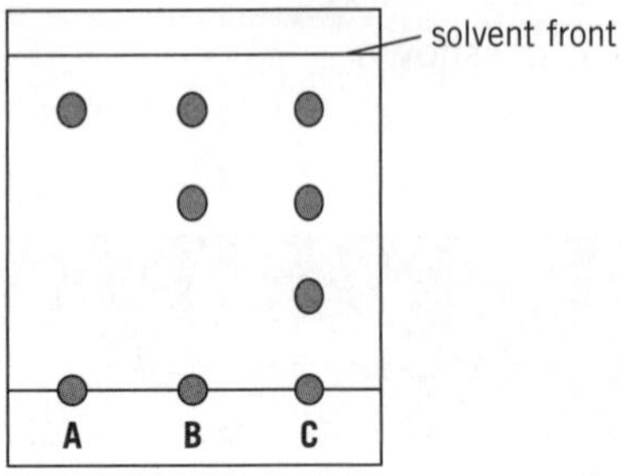

solvent front

A   B   C

Exam Tip

Draw horizontal lines across from each spot so you can see which other ink samples they appear in.

Write **two** conclusions that compare the dyes in ink samples **A**, **B**, and, **C**.   [2 marks]

06.4 Calculate the R_f value for the spot obtained from ink sample **A**.   [2 marks]

06.5 Circle the dye that is more attracted to the paper than all the others dyes. Justify your answer.   [2 marks]

07 A student had three unknown ionic compounds, **A**, **B**, and **C**. The student carried out tests on the compounds to identify them. **Table 5** shows their observations.

Table 5

Compound	Tests with solution			Test with solid
	Add sodium hydroxide solution	Add silver nitrate solution	Add barium chloride solution	Add dilute hydrochloric acid
A	blue precipitate	no change	no change	bubbles observed
B	no change	white precipitate	no change	no change
C	brown precipitate	no change	precipitate, colour difficult to see	no change

07.1 Name the acid that should be added with silver nitrate.   [1 mark]

07.2 Identify compound **A**. Justify your answer.   [3 marks]

07.3 Suggest a practical procedure the student could carry out to confirm the colour of the precipitate formed when barium chloride solution is added to compound **C**.   [1 mark]

07.4 Write a conclusion about the identity of compound **B**.   [2 marks]

07.5 Suggest a further test the student could carry out on compound **B** to help to identify the positive ion.   [1 mark]

07.6 Use your answer to **07.5** to give **three** possible observations and a conclusion for each one.   [3 marks]

Exam Tip

Start with identifying the metal ion and then the non metal ion.

08 **Figure 4** shows the flame emission spectra of six metal ions.

Figure 4

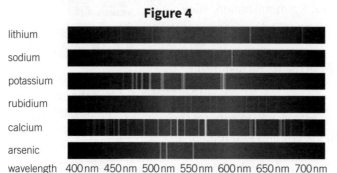

08.1 A chemist produced a flame emission spectrum from a solution containing two of the ions. The spectrum is shown in **Figure 5**.

Figure 5

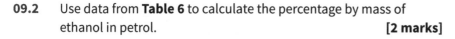
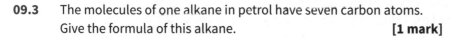

Name the **two** metal ions that are shown in the spectrum in **Figure 5**. **[1 mark]**

08.2 A student has a solution of lithium bromide. Describe **two** tests that the student could do to confirm the identity of the solution. Include the expected results in your answer. **[3 marks]**

08.3 Give **two** advantages of using instrumental methods to identify substances, compared to using chemical tests. **[2 marks]**

> **! Exam Tip**
> This question is only asking about advantages, writing disadvantages won't get you any marks.

09 Petrol is used to fuel cars. **Table 6** shows the different substances that are mixed in a sample of petrol.

Table 6

Substance	Mass of substance in 200 g of petrol in g
alkanes	110
other hydrocarbons	70
ethanol	20

09.1 Petrol is an example of a formulation. Define the term formulation. **[1 mark]**

09.2 Use data from **Table 6** to calculate the percentage by mass of ethanol in petrol. **[2 marks]**

> **! Exam Tip**
> Remember to use the general formula for alkanes.

09.3 The molecules of one alkane in petrol have seven carbon atoms. Give the formula of this alkane. **[1 mark]**

09.4 Ethanol in petrol is made from plants. The alkanes in petrol are obtained from crude oil. Suggest **one** advantage of including ethanol in petrol. **[1 mark]**

10 A firework technician wanted to identify the compound used in a firework. They knew the compound is a metal halide. They collected a sample of the compound and carried out a flame test. The flame turned green.

10.1 Identify the metal in the compound from the flame test result.

[1 mark]

Exam Tip

Go over the text and pick out the important bits of information before you start the question, this will means you can see what you need before starting the answers.

10.2 The technician used flame emission spectroscopy to confirm their result. **Figure 6** shows the technician's result and four reference spectra.

Identify the metal in the compound. [1 mark]

Figure 6

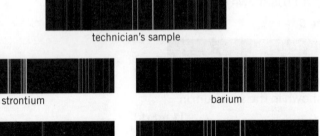

technician's sample

strontium

barium

potassium

copper

10.3 Suggest why the flame test may give the wrong answer.

[1 mark]

10.4 The technician then dissolved some of the compound in water. They added silver nitrate solution and nitric acid. A white precipitate formed. Identify the halide ion. [1 mark]

11 For each of the following pairs of substances, suggest **one** chemical test that you can use to tell them apart. Give the result of the test for both substances.

11.1 sodium carbonate and sodium nitrate [3 marks]

Exam Tip

There will need to be three tests and results for each pair.

11.2 potassium chloride and potassium iodide [3 marks]

11.3 calcium chloride and magnesium chloride [3 marks]

11.4 iron(II) sulfate and iron(III) sulfate [3 marks]

12 A student has a solution of an unknown aluminium compound. They only have a small sample of the solution, so they decide to carry out all three tests for the negative ion in the same test tube. The method they plan to use is:

1 Test for carbonate: add $1\,cm^3$ dilute sulfuric acid. Collect gas given off and pass through limewater.

2 Test for sulfates: add $1\,cm^3$ barium chloride and $1\,cm^3$ hydrochloric acid.

3 Test for halides: add $1\,cm^3$ nitric acid and $1\,cm^3$ of silver nitrate solution.

12.1 Explain why the suggested method will give a false result. **[2 marks]**

12.2 Describe how the method can be improved to prevent false results. **[1 mark]**

12.3 Suggest **one** improvement to the method that would reduce the number of substances needed to carry out the tests. **[1 mark]**

12.4 Write a balanced symbol equation for the reaction that occurs in the test for carbonates. Assume the unknown compound is aluminium carbonate, $Al_2(CO_3)_3$. **[3 marks]**

13 Companies crack long-chain hydrocarbons to produce more useful shorter-chain hydrocarbons. The equation shows an example of a cracking reaction.

$$C_{19}H\text{\_\_\_\_} \rightarrow C_8H_{20} + C\text{\_\_\_\_}H_{10} + C_6H_{10}$$

13.1 Complete the chemical formulae in the symbol equation. **[1 mark]**

13.2 Name the homologous series that the original compound belongs to. **[1 mark]**

13.3 Identify the product from the cracking reaction that is an alkane. **[1 mark]**

13.4 Identify the product from the cracking reaction that has two double bonds. **[1 mark]**

13.5 Complete the diagram to show the possible structure for the compound with two double bonds. **[3 marks]**

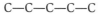

C—C—C—C—C

13.6 What are the conditions required for catalytic cracking? **[1 marks]**

> **! Exam Tip**
>
> Each carbon will only make four bonds in total, double bonds count as 2. Each hydrogen will only ever make one bond.

14 This question is about the following compounds

A C_3H_6 **C** CH_3CH_2COOH

B C_3H_8 **D** $CH_3CH_2CH_2OH$

14.1 Name compound **A**. **[1 mark]**

14.2 Name the homologous series that compound **D** belongs to. **[1 mark]**

14.3 Calculate the number of moles in 10.0 g of compound **C**. **[3 marks]**

14.4 Compare the chemical properties of compounds **C** and **D**. **[6 marks]**

> **! Exam Tip**
>
> Remember that the relative atom mass of these elements are:
>
> C = 12
>
> H = 1
>
> O = 16

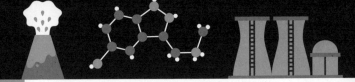

Knowledge

C18 The Earth's atmosphere

The Earth's changing atmosphere

The table below shows how the composition of the atmosphere has changed over the course of the Earth's entire 4.6 billion year history

Period	Proportions of gases	Evidence
about 4.6 billion years to about 2.7 billion years ago	• **carbon dioxide, CO_2** Released by volcanoes. Biggest component of the **atmosphere**. • **oxygen, O_2** Very little oxygen present. • **nitrogen, N_2** Released by volcanoes. • **water vapour, H_2O** Released by volcanoes. Existed as vapour as Earth was too hot for it to condense. • **other gases** Ammonia, NH_3, and methane, CH_4, may also have been present.	Because it was billions of years ago there is very little evidence to draw upon.

about 2.7 billion years ago to about 200 million years ago	• **carbon dioxide, CO_2**
	Amount in atmosphere begins to reduce because:
	  • water condenses to form the oceans, in which CO_2 then dissolves
	  • algae (and later plants) start to photosynthesise
	
	$$\text{carbon dioxide } + \text{ water } \xrightarrow{\text{light}} \text{ glucose } + \text{ oxygen}$$
	$$6CO_2 + 6H_2O \longrightarrow C_6H_{12}O_6 + 6O_2$$
	
	  • CO_2 precipitates in the oceans as solid carbonates (sediments) that form rocks
	  • CO_2 taken in by plants and animals. When they die, the carbon in them is locked up as fossil fuels
	• **oxygen, O_2**
	Starts to increase as a product of photosynthesis.
	• **nitrogen, N_2**
	Continues to increase. Nitrogen is a very stable molecule so any process that produces it causes the overall amount to build up over time.
	• **water vapour, H_2O**
	Starts to decrease. As the Earth cools, the vapour condenses and forms the oceans.
		Still limited as billions of years ago, but can look at processes that happen today (like photosynthesis) and make theories about the past.

about 200 million years ago until the present	• **carbon dioxide, CO_2** about 0.04% • **oxygen, O_2** about 20% • **nitrogen, N_2** about 80% • **water vapour, H_2O** Very little overall. Collects in large clouds as part of the water cycle. • **other gases** Small proportions of other gases such as the noble gases.	Ice core evidence for millions of years ago and lots of global measurements taken recently.

small proportions of other gases, such as water vapour, carbon dioxide, and noble gases

oxygen ~20%

nitrogen ~80%

Key terms

Make sure you can write a definition for these key terms.

acid rain   atmosphere   carbon footprint   global climate change

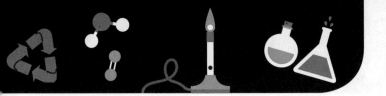
The information on this page shows how very recent human activity (in the past 150 years or so) is increasing the amount of CO_2 in the atmosphere.

Greenhouse gases

Greenhouse gases, such as carbon dioxide, methane, and water vapour, absorb radiation and maintain temperatures on the Earth to support life.

However, in the last 150 years, more greenhouse gases have been released due to human activities.

- carbon dioxide – combustion of fossil fuels, deforestation
- methane – planting rice fields, cattle farming

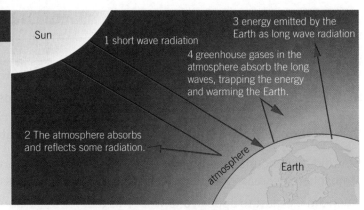

Sun

1 short wave radiation

2 The atmosphere absorbs and reflects some radiation.

3 energy emitted by the Earth as long wave radiation

4 greenhouse gases in the atmosphere absorb the long waves, trapping the energy and warming the Earth.

atmosphere   Earth

Global warming

Scientists have gathered peer-reviewed evidence to demonstrate that increasing the amount of greenhouse gases in the atmosphere will increase the overall average temperature of the Earth. This is called **global warming**.

However, it is difficult to make predictions about the atmosphere as it is so big and complex. This leads some people to doubt what scientists say.

Global climate change

Global warming leads to another process called **global climate change** – how the overall weather patterns over many years and across the entire planet will change.

There are many different effects of climate change, including:

- sea levels rising
- extreme weather events
- changes in the amount and time of rainfall
- changes to ecosystems and habitats
- polar ice caps melting.

Carbon footprints

Increasing the amount of greenhouse gases in the atmosphere increases the global average temperature of the Earth, which results in global climate change.

As such, it is important to reduce the release of greenhouse gases into the atmosphere. The amount of carbon dioxide and methane that is released into the atmosphere by a product, person, or process is called its **carbon footprint**.

Other pollutants released in combustion of fuels

Pollutant	Origin	Effect
carbon monoxide	incomplete combustion of fuels	colourless and odourless toxic gas
particulates (soot and unburnt hydrocarbons)	incomplete combustion of fuels especially in diesel engines	**global dimming**, respiratory problems, potential to cause cancer
sulfur dioxide	sulfur impurities in the fuel reacting with oxygen from the air	**acid rain** and respiratory problems
oxides of nitrogen	nitrogen from the air being heated near an engine and reacting with oxygen	acid rain and respiratory problems

global dimming   global warming   greenhouse gas   particulate   pollutant

Learn the answers to the questions below then cover the answers column with a piece of paper and write as many as you can. Check and repeat.

	C18 questions		Answers
1	What is the atmosphere?	Put paper here	a layer of gas surrounding the Earth
2	What was the early atmosphere composed of?		mostly carbon dioxide
3	How did the oceans form?		water vapour condensing as the Earth cooled
4	How did the amount of carbon dioxide in the atmosphere decrease to today's levels?		dissolved in the oceans, photosynthesis, converted to fossil fuels, precipitated as insoluble metal carbonates
5	When did life start to appear, and what was the impact of this on oxygen in the atmosphere?		about 2.7 billion years ago; amount of atmospheric oxygen increased as it was released in photosynthesis
6	How has the amount of nitrogen in the atmosphere changed over time?	Put paper here	increased slowly as it is a very stable molecule
7	Why can scientists not be sure about the composition of the Earth's early atmosphere?		it was billions of years ago and evidence is limited
8	What is the current composition of the atmosphere?		approximately 80% nitrogen, 20% oxygen, and trace amounts of other gases such as carbon dioxide, water vapour, and noble gases
9	What is a greenhouse gas?		a gas that traps radiation from the Sun
10	What type of radiation do greenhouse gases absorb?	Put paper here	longer wavelength infrared radiation
11	Name three greenhouse gases.		methane, carbon dioxide, water vapour
12	Give two ways recent human activities have increased the amount of atmospheric carbon dioxide.		burning fossil fuels, deforestation
13	Give two ways recent human activities have increased the amount of atmospheric methane.		rice farming, cattle farming
14	What is global warming?		an increase in the overall global average temperature
15	What is global climate change?	Put paper here	the change in long-term weather patterns across the planet
16	What are some possible effects of climate change?		sea levels rising, extreme weather events, changes in the amount and time of rainfall, changes to ecosystems and habitats, polar ice caps melting
17	What is a carbon footprint?	Put paper here	the amount of carbon a product, process, or person releases into the atmosphere over its lifetime
18	How is carbon monoxide formed, and what is the danger associated with it?		incomplete combustion; colourless and odourless toxic gas
19	How are particulates formed, and what are the dangers associated with them?		incomplete combustion; global dimming, respiratory problems, potential to cause cancer
20	How is sulfur dioxide formed, and what are the dangers associated with it?		sulfur impurities in fossil fuels react with oxygen during combustion; acid rain, respiratory problems
21	How are oxides of nitrogen formed, and what are the dangers associated with them?		atmospheric oxygen and nitrogen react in the heat of a combustion engine; acid rain, respiratory problems

Now go back and use the questions below to check your knowledge from previous chapters.

C18

Previous questions | Answers

#	Previous questions		Answers
1	What are the two types of polymerisation?	Put paper here	addition and condensation
2	In terms of electrons, what is reduction?		gain of electrons
3	What is the name of the positive electrode?		anode
4	What is a strong acid?	Put paper here	an acid where the molecules or ions completely ionise in water
5	What is bond energy?		the energy required to break a bond or the energy released when a bond is formed
6	What are some examples of formulations?	Put paper here	fuels, cleaning agents, paints, medicines, alloys, fertilisers, and foods
7	What are amino acids?		the building blocks for polypeptides and proteins, which have an amine and a carboxylic acid group
8	What is the general formula for alkenes?		C_nH_{2n}

Maths Skills

Practise your maths skills using the worked example and practice questions below.

Lines of best fit	Worked Example	Practice
When describing lines of best fit, you need to state: • its correlation • if the line is straight or curved • whether the line plateaus (stops changing, and flattens out) • whether the line runs through the origin (0,0). Correlations can be positive or negative and either strong or weak, or there can be no correlation. If the line of best fit is straight and goes through the origin, the variables are **directly proportional** to each other, meaning as one variable changes the other changes at the same rate.	Fully describe the curved line of best fit for the reaction plotted on the graph below. 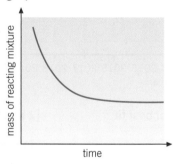 **Answer:** The graph shows a negative correlation. The curved line of best fit does not pass through the origin. The mass of the mixture decreases rapidly at first, this decrease then slows down in the middle of the reaction, and finally plateaus as the mass stops decreasing with time.	The graph below shows how the volume of gas produced changes with time in the reaction between marble chips and hydrochloric acid. 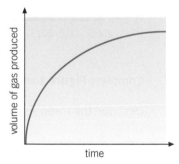 1 Describe the graph. 2 Sketch a graph to show a directly proportional relationship between two variables.

Exam-style questions

01 **Table 1** and **Figure 1** show the average concentration of carbon dioxide in the atmosphere every January from 2010 to 2019.

Table 1

Year	Average concentration of CO_2 in January in parts per million
2010	389
2011	391
2012	393
2013	395
2014	398
2015	400
2016	403
2017	406
2018	408
2019	411

Figure 1

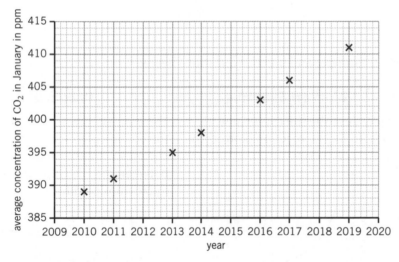

01.1 Complete **Figure 1** and draw a line of best fit. **[2 marks]**

01.2 Describe the trend shown by the graph. **[1 mark]**

_____

_____

01.3 Describe how **two** human activities are responsible for the trend shown in **Figure 1**. **[4 marks]**

1 _____

_____

_____

2 _____

_____

_____

> **! Exam Tip**
>
> Place a clear ruler over the points and see where most of them fit the line and an even number is on either side.

> **! Exam Tip**
>
> This is only a 1 mark question, so a simple description of the shape is all that is needed.

02   This question is about air pollutants.

02.1   Which pollutant is formed in car engines from the reaction between two gases that occur naturally in the atmosphere?   **[1 mark]**

Tick **one** box.

carbon dioxide [   ]   oxides of nitrogen [   ]

carbon particles [   ]   sulfur dioxide [   ]

> **! Exam Tip**
> Think about which gases are found at the highest levels in the atmosphere.

02.2   Carbon monoxide is also produced in car engines.
Name the process that produces carbon monoxide.   **[1 mark]**

_____

02.3   Balance the symbol equation for the incomplete combustion of a fuel.   **[1 mark]**

____ $C_4H_{10}(g)$ + ____ $O_2(g)$ → ____ $CO(g)$ + ____ $H_2O(l)$

> **! Exam Tip**
> Start with the carbons, then the hydrogens and leave the oxygens until last!

02.4   Draw **one** line from each pollutant to an effect of the pollutant.   **[3 marks]**

Pollutant   **Effect**

oxides of nitrogen	poisoning of humans
	global dimming
carbon monoxide	global climate change
particulates	breathing problems

> **! Exam Tip**
> Not all of the boxes on the right hand side will be used.

02.5   Which gas causes acid rain?   **[1 mark]**
Tick **one** box.

carbon dioxide [   ]   sulfur dioxide [   ]

carbon monoxide [   ]   unburnt hydrocarbons [   ]

03 An international organisation suggests three ways in which to reduce the rate of global climate change. Evaluate each suggestion (**03.1**, **03.2**, and **03.3**) in terms of how effective it would be in reducing global climate change, and what socioeconomic effects it could have.

03.1 All governments invest in alternatives to fossil fuels. **[4 marks]**

03.2 Stop South American countries from cutting down the rainforest for farmland. **[3 marks]**

03.3 Tax cars to encourage use of other modes of transport. **[4 marks]**

! **Exam Tip**

For an evaluate question, you need good points, bad points, *and* your opinion with a reason.

04 **Figure 2** shows the gases that make up the atmosphere of the planet Mars.

04.1 Compare the composition of the atmosphere of Mars to the composition of the atmosphere of the Earth. **[6 marks]**

04.2 Methane gas is also present in the atmosphere of Mars. The concentration of methane was measured at 0.7 parts per billion in August 2013. In June 2019 it was measured again at 21.0 parts per billion. How many times greater was the concentration of methane in June 2019 compared to August 2013? **[1 mark]**

04.3 Suggest why scientists cannot be certain about the reasons for the changes in methane concentration in the atmosphere of Mars. **[1 mark]**

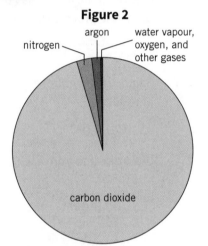

Figure 2

nitrogen — argon — water vapour, oxygen, and other gases

carbon dioxide

04.4 The average surface temperature of Mars is around −60 °C. The atmosphere of Mars is about 100 times less dense than the atmosphere of Earth. Explain why the surface temperature of Mars is significantly lower than on Earth. **[4 marks]**

04.5 During the Martian winter, carbon dioxide in the atmosphere condenses to form polar ice caps made from carbon dioxide. During the summer, this carbon dioxide transforms back into a gas and returns to the atmosphere. Name the process by which carbon dioxide changes state from a solid to a gas. **[1 mark]**

! **Exam Tip**

For question **04.1**, compare means you need to give similarities *and* differences between the two planets.

05 The Earth's atmosphere contains oxygen gas.

05.1 Draw a dot and cross diagram to show the bonding in an oxygen molecule. **[2 marks]**

05.2 Explain why the percentage of oxygen in the atmosphere changed from about 2.7 billion years ago to the present. Include a balanced symbol equation in your answer. **[6 marks]**

! **Exam Tip**

Don't forget oxygen gas is diatomic molecule with a double bond between the oxygen atoms.

05.3 Suggest how planting trees could help reduce the effects of global climate change. **[3 marks]**

05.4 Explain how acid rain can contribute to global climate change. **[2 marks]**

06 Carbon dioxide is a greenhouse gas.

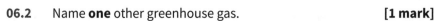

06.1 What is meant by the term greenhouse gas? **[1 mark]**

06.2 Name **one** other greenhouse gas. **[1 mark]**

06.3 Give **two** human activities that increase the amount of carbon dioxide in the atmosphere. **[2 marks]**

06.4 Increasing amounts of greenhouse gases result in an increase in average global temperature. This is a major cause of climate change. Give **three** effects of global climate change. **[3 marks]**

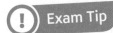

> **! Exam Tip**
>
> **06.4** is worth 3 marks, so three short effects are all thats needed and not long explanations of each effect.

07 This question is about the carbon footprint of a journey by road. **Table 2** shows carbon dioxide emissions data for a car and a bus.

07.1 A car travels 120 km. Use data from **Table 2** to calculate the mass of CO_2 emitted by the car on this journey. Give your answer in kg. **[2 marks]**

07.2 Two people are travelling in the car during the 120 km journey. Calculate the mass of CO_2 emitted by the car per person for this journey. Give your answer in kg. **[1 mark]**

07.3 A bus travels on the same 120 km journey as the car. Calculate the minimum number of people that must be on the bus in order for the mass of CO_2 emitted per person to be **less than** the mass emitted for two people travelling the same journey in the car. **[3 marks]**

07.4 Suggest why some people do not travel by bus instead of by car, even though CO_2 emissions can be smaller by bus. **[1 mark]**

Table 2

Vehicle	Mass of CO_2 emitted by the vehicle in grams per kilometre
car	100
bus	1050

> **! Exam Tip**
>
> Don't forget to convert from grams to kg.

08 **Figure 3** shows values for the global annual mean surface temperature change between 2000 and 2018.

Figure 3

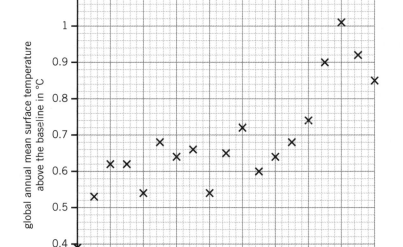

08.1 Give the temperature above the baseline in 2005. **[1 mark]**

08.2 Give the **four** years with the highest temperature. **[1 mark]**

08.3 Give the year in which the global annual mean surface temperature above the baseline was 0.53 °C. **[1 mark]**

08.4 Describe the overall trend shown by **Figure 3**. **[1 mark]**

08.5 Explain how human activities have contributed to the trend shown in **Figure 3**. **[2 marks]**

08.5 A student says that **Figure 3** shows that global temperature is now decreasing. Evaluate this statement. **[3 marks]**

09 A healthy diet must include protein. **Table 3** gives the carbon footprint for the production of some foods that are high in protein.

09.1 Suggest why data for food carbon footprints is published. **[1 mark]**

09.2 People that follow a vegetarian diet do not eat meat. Evaluate swapping from a meat-based diet to a vegetarian diet as a way of reducing an individual's carbon footprint. **[3 marks]**

09.3 Use data from **Table 3** to calculate the total mass of CO_2 emissions when 50 g of beef and 25 g of cheese are produced. **[2 marks]**

09.4 Beef and cheese are produced from cattle. Suggest why **Table 3** is not an accurate representation of the carbon footprint of these foods. **[1 mark]**

Table 3

Food	Typical CO_2 emissions to produce 100 g of food in kg
beef	16
cheese	12
chicken	7
egg	5
nuts	1
beans and peas	1

10 The composition of the Earth's atmosphere has changed since the Earth formed about 4.6 billion years ago.

10.1 Why is there little evidence about the early atmosphere? **[1 mark]**

10.2 Explain how the percentages of carbon dioxide and oxygen in the atmosphere have changed over the past 4.6 billion years. **[6 marks]**

10.3 Name **two** greenhouse gases that are added into the atmosphere as a result of human activity. **[2 marks]**

10.4 Give **one** effect of human activity increasing the greenhouse gases in the atmosphere. **[1 mark]**

11 Carbon dioxide, methane, and water vapour all act as a greenhouse gas in the atmosphere.

11.1 Explain how greenhouse gases maintain the temperature of the Earth. **[4 marks]**

11.2 The combustion of fossil fuels produces carbon dioxide. Describe **one** other way in which burning fossil fuels can contribute to an increase in the percentage of carbon dioxide in the atmosphere. **[3 marks]**

Exam Tip

Don't just repeat the example given in the question.

11.3 The higher the temperature of the atmosphere, the more water vapour the atmosphere is able to hold. Explain how human activities may lead to an increase in water vapour in the atmosphere, and give the effect this could have on global climate change. **[5 marks]**

12 A student has three metal compounds labelled **A**, **B**, and **C**. The student knows that they are a calcium compound, iron(II) sulfate, and an aluminium compound, but they do not know which is which.

12.1 Describe how the student can identify which metal compound is which. **[5 marks]**

12.2 The student identified compound **B** as the iron compound. Describe how the student can confirm that compound **B** is iron sulfate. **[3 marks]**

12.3 Write a balanced symbol equation for the reaction to identify the iron(II) in **12.1**. **[2 marks]**

12.4 The student identified that compound **A** was the calcium compound and that compound **C** was the aluminium compound. The student tested both compounds with silver nitrate solution. Compound **A** produced a yellow precipitate. Compound **C** produced a white precipitate. Identify the **two** compounds. **[2 marks]**

13 **Figure 4** shows a reaction profile.

13.1 Write the label that should be on the y-axis. **[1 mark]**

13.2 Give the letter of the arrow that shows the activation energy of the reaction. **[1 mark]**

13.3 Explain whether the reaction is exothermic or endothermic. **[2 marks]**

13.4 Give **one** example of the type of reaction from **13.3**. **[1 mark]**

Figure 4

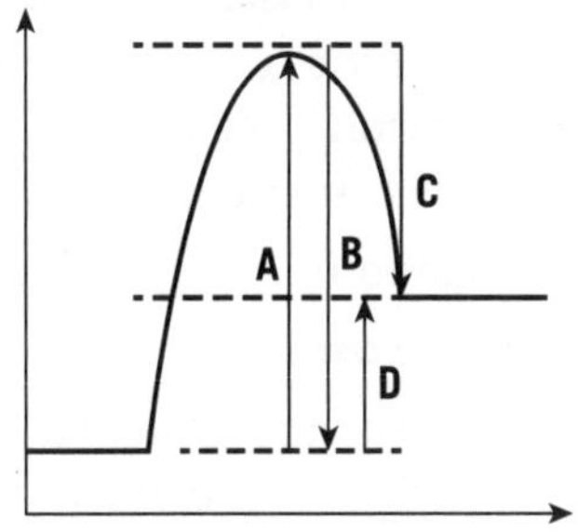

14 This question is about ethane, C_2H_6, and ethene, C_2H_4.

14.1 Draw a dot and cross diagram to show the bonds in ethene. **[3 marks]**

14.2 **Table 4** gives the energy needed to break single and double carbon–carbon bonds.

Give a reason for the greater strength of the carbon–carbon double bond. **[1 mark]**

Table 4

Bond	Bond energy in kJ/mol
C—C	348
C=C	642

14.3 Suggest why ethene takes part in more reactions than ethane, even though ethene has the stronger carbon–carbon bond. **[2 marks]**

Knowledge

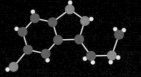

C19 Using the Earth's resources A

Natural and synthetic resources

We use the Earth's resources to provide us with warmth, fuel, shelter, food, and transport.

- Natural resources are used for food, timber, clothing, and fuels.
- Synthetic resources are made by scientists. They can replace or supplement natural resources.

When choosing and synthesising resources, it is important to consider **sustainable development**. This is development that meets the needs of current generations without compromising the ability of future generations to meet their own needs.

Finite and renewable resources

Some resources are **finite**. This means that they will eventually run out.

Fossil fuels are an example of a finite resource. They take so long to form that we use them faster than they are naturally formed.

Resources that will not run out are called **renewable** resources.

Wood is an example of a renewable resource. Trees can be grown to replace any that are cut down for wood.

Potable water

Water is a vital resource for life. **Potable** water is water that is safe to drink. However, most water on Earth is not potable.

Type of water	What it has in it
pure water	just water molecules and nothing else
potable water	water molecules, low levels of salts, safe levels of harmful microbes
salty water (sea water)	water molecules, dangerously high levels of salt, can have high levels of harmful microbes
fresh water (from rivers, lakes, or underground)	water molecules, low levels of salt, often has harmful microbes at high levels

Fresh water

In the UK, potable water is produced from rain water that collects in lakes and rivers. To produce potable water:

1 Choose an appropriate source of fresh water.
2 Pass the water through filters to remove large objects.
3 **Sterilise** the water to kill any microbes using ozone, chlorine, or UV light.

Salty water

Some countries do not have lots of fresh water available. **Desalination** is the process to turn saltwater into potable water. This requires a lot of energy and can be done by:

- distillation
- **reverse osmosis**

Reverse osmosis involves using membranes to separate the salts dissolved in the water. The water needs to be pressurised and the salty water corrodes the pumps. As such, it is an expensive process.

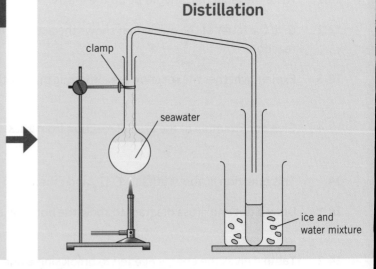

Distillation

Key terms

Make sure you can write a definition for these key terms.

aerobic anaerobic distillation effluent finite resources potable water renewable resources reverse osmosis screening sedimentation sewage sludge sterilisation sustainable development

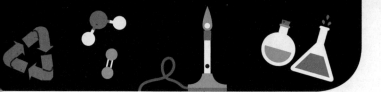

Waste water

Human activities produce lots of waste water as sewage, agricultural waste, and industrial waste.

- **Sewage** and agricultural waste contain organic matter and harmful microbes.
- Industrial waste contains organic matter and harmful chemicals.

These need to be removed before the water can be put back into the environment.

Treating sewage water

screening and grit removal
The sewage passes through a metal grid that filters out large objects.

sedimentation
The sewage is left so that solid sediments settle out of the water. The sediments sink to the bottom of the tank. The liquid sits above the sediment.

Treating sludge

sewage sludge
This sediment is called **sludge**. Sludge contains organic matter, water, dissolved compounds, and small solid particles.

anaerobic treatment
Bacteria are added to digest the organic matter. These bacteria break down the matter anaerobically – with a limited supply of oxygen.

biogas
The anaerobic digestion of sludge produces biogas. Biogas is a mixture of methane, carbon dioxide and hydrogen sulfide. It can be used as fuel.

remaining sludge used as fuel
The remaining sludge can be dried out and can also be burnt as a fuel.

Treating effluent

effluent
The remaining liquid is called **effluent**. This effluent has no solid matter visible, but still contains some matter and harmful microorganisms.

aerobic treatment
Bacteria are added to the effluent. These bacteria feed on organic matter and the harmful microorganisms in the effluent. The bacteria break down the matter by aerobic respiration – oxygen needs to be present.

bacteria removed
The bacteria are allowed to settle out of the water.

discharged back to rivers
The water is now safe enough to be released back into the environment.

Knowledge

 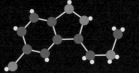 

C19 Using the Earth's resources B

Metal extraction

Metals are used for many different things. Some metals can be extracted from their ores by reduction or electrolysis.

However, metal ores are a finite resource and these processes require lots of energy.

Scientists are looking for new ways to extract metals that are more sustainable.

Phytomining and **bioleaching** are two alternative processes used to extract copper from low grade ores (ores with only a little copper in them).

Phytomining

1 Grow plants near the metal ore.
2 Harvest and burn the plants.
3 The ash contains the metal compound.
4 Process the ash by electrolysis or displacement with scrap metal.

Bioleaching

1 Grow bacteria near the metal ore.
2 Bacteria produce leachate solutions that contain metal compound.
3 Process the leachate by electrolysis or displacement with scrap metal.

Both of these methods avoid the digging, moving, and disposing of large amounts of rock associated with traditional mining techniques.

Life cycle assessment

A **life cycle assessment (LCA)** is a way of looking at the whole life of a product and assessing its impact on the environment and sustainability. It is broken down into four categories:

• extracting and processing raw materials
• manufacturing and packaging
• use and operation during its lifetime
• disposal at the end of its useful life, including transport and distribution at each stage

Some parts of an LCA are objective, such as the amount of water used or waste produced in the production of a product.

However, other parts of an LCA require judgements, such as the polluting effect, and so LCAs are not a completely objective process.

 **Key terms**

Make sure you can write a definition for these key terms.

| biodegrade | bioleaching | life cycle assessment | phytomining | recycling |

C19

Disposal of products

When someone finishes with a product, it can be

- added to a landfill
This can cause habitat loss and other problems in the local ecosystem. Some items persist in landfills as they do not **biodegrade** and could be there for hundreds of years.

- incinerated
Some products can be incinerated to produce useful energy. However, the combustion can often be incomplete and result in harmful pollutants being released to the atmosphere.

- reused
This is when an item is used again for a similar purpose.

- **recycled**
Recycling requires energy, but conserves the limited resources and often requires less energy than needed to make brand new materials.

The table shows information about the extraction, processing, and disposal of some common materials.
This information is used when making a LCA.

Material	Extraction/processing	Disposal
metal	• quarrying and mining cause habitat loss • machinery involved in mining release greenhouse gases • extraction from metal ores require lots of energy	• metals can normally be recycled by melting them down and then casting them into new shapes • metals in landfill can persist for a long time
plastic	normally come from fossil fuels that are non-renewable	• many plastic products can be reused and recycled • plastics often end up in landfills where they persist as they are not biodegradable • incinerating plastics releases lots of harmful pollutants like carbon monoxide and particulates
paper	produced from trees that require land and lots of water to grow lots of water also used in the production process	• many paper products can be recycled • paper products can also be incinerated or they can decay naturally in a landfill • incineration and decay release greenhouse gases
glass	produced by heating sand, which requires a lot of energy	• many glass products can be reused, or crushed and recycled • if glass is added to landfills it persists as it is not biodegradable
ceramics	• come from clay and rocks • generally require quarrying, which requires energy, releases pollutants from heavy machinery, and causes habitat loss	• most ceramics are not commonly recycled in the UK, and once broken cannot be reused • ceramics tend to persist in landfills

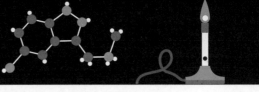

Learn the answers to the questions below then cover the answers column with a piece of paper and write down as many as you can. Check and repeat.

C19 questions | Answers

	C19 questions		Answers
1	What do we use the Earth's resources for?		warmth, shelter, food, fuel, transport
2	What are some examples of natural resources?		cotton, wool, timber
3	What are some examples of synthetic resources?		plastic, polyester, acrylic
4	What is a finite resource?		a resource that will eventually run out
5	What is sustainable development?		development that meets the needs of current generations without compromising the ability of future generations to meet their own needs
6	What are the four main types of water?		pure water, salt water, fresh water, potable water
7	What is potable water?		water that is safe to drink
8	In the UK, how is potable water extracted from fresh water?		filtration and sterilisation
9	What is sterilisation?		killing microbes
10	What are three examples of sterilising agents?		chlorine gas, UV light, and ozone
11	How can potable water be produced from salt water?		desalination
12	How can desalination be carried out?		distillation or reverse osmosis
13	What are the three main types of waste water?		sewage, agricultural waste, industrial waste
14	What can waste water contain?		organic matter, harmful microbes, harmful chemicals
15	What is the first step in processing waste water?		screening and grit removal
16	What is sedimentation?		separating the waste water into sludge and effluent
17	How is sludge treated?		anaerobic respiration
18	How is effluent treated?		aerobic respiration
19	What is phytomining?		using plants to extract copper
20	What is bioleaching?		using bacteria to extract copper
21	What is a life cycle assessment?		a way of assessing the energy costs and environmental effect of a product across its lifetime
22	What are the four stages of a life cycle assessment?		• extracting and processing raw materials • manufacturing and packaging • use and operation during its lifetime • disposal at the end of its useful life
23	How can we reduce the amount of new materials manufactured?		by reducing, reusing, or recycling products
24	In what ways can materials that are not recycled be disposed?		landfill or incineration

Put paper here

Now go back and use the questions below to check your knowledge from previous chapters.

Previous questions | Answers

	Previous questions		Answers
1	What effect does increasing temperature have on the rate of reaction?	Put paper here	increases
2	Why does increasing temperature have this effect?	Put paper here	particles move faster, leading to more frequent collisions – particles have the same activation energy, so more collisions result in a reaction
3	What information does flame emission spectroscopy produce?	Put paper here	the wavelength of light given off by a metal in a flame to identity of the metal and its concentration
4	What is a greenhouse gas?	Put paper here	a gas that traps radiation from the Sun
5	How is R_f calculated?	Put paper here	$R_f = \dfrac{\text{distance moved by substance}}{\text{distance moved by solvent}}$
6	What colour precipitates are formed when sodium hydroxide solution is added to solutions of copper(II), iron(II), and iron(III) ions?	Put paper here	copper(II) ions form blue precipitate, iron(II) ions form green precipitate, iron(III) ions form brown precipitate
7	What is the effect of increasing the concentration of reactants on a reaction at dynamic equilibrium?	Put paper here	favours the forward reaction
8	When did life start to appear, and what was the impact of this on oxygen in the atmosphere?	Put paper here	about 2.7 billion years ago; amount of atmospheric oxygen increased as it was released in photosynthesis

Required Practical Skills

Practise answering questions on the required practicals using the example below.
You need to be able to apply your skills and knowledge to other practicals too.

Water purification	Worked Example	Practice
You need to be able to describe how to analyse the purity of a water sample, and how to use distillation to purify the sample. To do this, you need to know how to test pH, and describe the method of distillation for any solution. In an exam, you may also be asked about the purity and purification of different samples other than water. You should also learn the different terms describing how water is made safe to drink.	A student wanted to determine the identity of a salt dissolved in a sample of water. They evaporated away $100\,cm^3$ of the 1 M solution. The empty evaporating basin weighed 92.78 g, and the basin containing the solids after evaporation weighed 98.63 g. Suggest how you could determine the identity of the salt. **Answer**: mass of solid salt = 98.63 – 92.78 = 5.85 g 5.85 g salt in $100\,cm^3$ = 58.5 g in $1\,dm^3$ $M_r = \dfrac{\text{mass}}{\text{moles}}$ $M_r = \dfrac{58.5}{1} = 58.5$ The salt has an M_r of 58.5, so use the relative atomic masses on the Periodic Table to determine a potential identity: 23 (Na) + 35.5 (Cl) = 58.5, so the salt could be NaCl. This could be confirmed using a flame test and a halide test.	1 Explain how you could use pH to determine if a sample of water is pure. 2 After carrying out a distillation experiment, a student re-distilled the distillate. Suggest what the student would have observed. 3 Describe the difference between pure and potable water.

 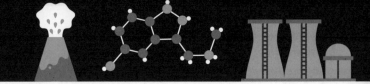

Exam-style questions

01   This question is about water treatment.

01.1   What is potable water?   **[1 mark]**
Tick **one** box.

water that has nothing
mixed with it   ☐

water that is pure   ☐

water that salt has been
removed from   ☐

water that is safe to drink   ☐

01.2   Draw **one** line from each water treatment process to the reason for
the process.   **[2 marks]**

Water treatment process	Reason for process
passing water through filter beds	kill microorganisms
sterilising	remove dissolved salts
desalination	remove pieces of solid

01.3   Name **two** sterilising agents that are used to make water safe
to drink.   **[2 marks]**

1 _____

2 _____

01.4   Give **one** advantage and **one** disadvantage of obtaining drinking
water by desalination.   **[2 marks]**

advantage: _____

_____

disadvantage: _____

_____

02   A student has an empty glass jar.

02.1   Which is an example of recycling?   **[1 mark]**
Tick **one** box.

washing the jar and using it to store hairclips   ☐

crushing and melting the jar to make a bottle   ☐

putting the jar in landfill   ☐

making an identical jar from raw materials   ☐

02.2   **Table 1** lists the raw materials that are used to make glass.

Table 1

Raw material	Chemical formula
silicon dioxide	
calcium carbonate	
	Na_2CO_3

Complete **Table 1**.   **[3 marks]**

> **! Exam Tip**
>
> It is important that you have learnt the key ions from the specification. You can use the charged on the ions to work out chemical formula of ionic copounds. This is a skill that will come up throughout the chemistry course.

02.3   Outline **three** advantages of making glass objects from recycled glass, rather than from glass that has been newly made from its raw materials.   **[3 marks]**

1 _____

_____

2 _____

_____

3 _____

_____

03   Some students are investigating water from different sources. They want to compare the mass of dissolved solids in three samples of water. They use the following method:

1   Find the mass of an empty evaporating basin.

2   Use a measuring cylinder to measure 10.0 cm³ of one of the water samples into the evaporating basin.

3   Heat the evaporating basin and its contents until all the water has evaporated.

4   Find the mass of the evaporating basin again.

> **! Exam Tip**
>
> Think about accuracy and safety.

For answers and more practice questions visit
www.oxfordrevise.com/scienceanswers   Even more practice and interactive
revision quizzes are available on    C19 Practice   201

03.1 Suggest an improvement to step **2**. Give a reason for this improvement. **[2 marks]**

03.2 Suggest an improvement to step **3**. Give a reason for this improvement. **[2 marks]**

03.3 Give **two** safety precautions that the students should take. **[2 marks]**

03.4 **Table 2** shows the students' results. The mass of the empty evaporating basin was 95.24 g.

Table 2

Water sample	Mass of solid and evaporating basin after heating in grams
A	95.24
B	95.26
C	95.61

Identify the water sample that is pure water. **[1 mark]**

03.5 Identify the water sample that is most likely to be seawater. Give a reason for your decision. **[2 marks]**

04 A student is using distillation to purify a sample of seawater. **Figure 1** shows the apparatus the student used.

Figure 1

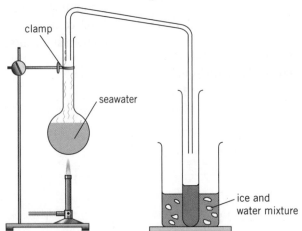

clamp

seawater

ice and water mixture

04.1 Give a reason for using a Bunsen burner, instead of a water bath, to heat the seawater. **[1 mark]**

04.2 Suggest a reason for using an ice–water mixture, not lumps of ice alone. **[1 mark]**

04.3 Identify the mistake the student has made in setting up the apparatus. **[1 mark]**

04.4 The student corrects the mistake, and starts the experiment again. Explain how the concentration of salt in the seawater changes as pure water is collected in the test tube. **[2 marks]**

05 Copper exists naturally on the Earth chemically bonded to non-metals. The pure metal can be extracted from these compounds.

05.1 Describe how copper is obtained from low-grade copper ore by phytomining. **[4 marks]**

05.2 Evaluate the advantages and disadvantages of obtaining copper by recycling scrap copper and bioleaching. **[6 marks]**

05.3 An ore of copper contains 22.1% copper. Calculate the mass of waste produced when 50.0 kg of copper is extracted from the ore. **[4 marks]**

06 Disposable water bottles are made from different materials. **Table 3** shows information on the life cycle assessments (LCAs) of two types of disposable water bottle. All quantities given are for the production of 12 bottles.

Table 3

Bottle material	PLA bottles	PET (plastic) bottles
Raw material	starch from plants	oil
Relative soil pollution	52.4	31.0
Land required in m²	0.234	0.0565
Global warming in kg of CO_2 equivalent	3.58	3.87
Energy used in production in megajoules	62.1	69.4
Is the material biodegradable?	yes, but the process is slow and can only occur if the conditions are correct	no
Is the material recyclable?	no	yes

06.1 Evaluate the use of PLA compared with PET to make disposable water bottles. **[6 marks]**

06.2 Calculate the area of land required to make **one** PET bottle. Give your answer in standard form to three significant figures. **[3 marks]**

06.3 Disposable water bottles can also be made from recycled PET. Predict how the energy used in production compares for recycled PET bottles and ordinary PET bottles. Suggest a reason for your prediction. **[2 marks]**

07 The steps below describe life cycle of a cotton T-shirt.

1 plant, grow and harvest cotton plants

2 make cotton thread and fabric

3 sew T-shirt

4 put into plastic bags

5 transport to buyer

6 buyer wears T-shirt regularly

7 one year later, buyer puts T-shirt in dustbin, which is taken to landfill

07.1 Identify the **three** steps that are part of the manufacturing and packaging stage of the life cycle of the T-shirt. **[1 mark]**

07.2 Life cycle assessments assess the impact of products in four stages. Give the four stages of a life cycle assessment. Identify **one** step in the lifetime of a cotton T-shirt for each stage in the life cycle assessment. **[4 marks]**

07.3 There are carbon dioxide emissions associated with the T-shirt when it is being used. Suggest why. **[1 mark]**

07.4 Suggest **one** action the buyer could take to reduce the environmental impact of the T-shirt at the disposal stage. **[1 mark]**

08 This question is about sewage treatment.

08.1 Name **two** types of substance that must be removed from household sewage and agricultural waste water, before the water is released back into the environment. **[2 marks]**

08.2 Give the **four** steps in the treatment of sewage. **[4 marks]**

08.3 Suggest why it is easier to obtain potable water from ground water than from sewage. **[1 mark]**

09 The metal tantalum is vital for making devices such as mobile phones. **Figure 2** shows the amounts of tantalum produced by different countries in one year.

Figure 2

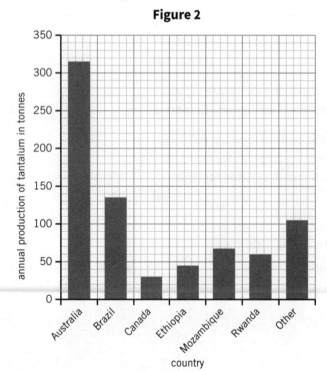

09.1 Calculate the total mass of tantalum produced. **[1 mark]**

09.2 Calculate the percentage of tantalum produced in Ethiopia. **[2 marks]**

09.3 A tantalum ore contains 72 % Ta_2O_5. Calculate the mass of tantalum in 80 kg of this ore. Relative atomic masses A_r: Ta = 181; O = 16 **[3 marks]**

09.4 The reserves of tantalum are the ore deposits that have been discovered but not yet mined. Worldwide known reserves of tantalum are approximately 1.5×10^5 tonnes. Estimate the year when known reserves of tantalum will run out, assuming annual production continues at the same rate as in **Figure 2**. **[2 marks]**

09.5 Suggest **two** reasons why tantalum might still be produced after the year given in **09.5**. **[2 marks]**

10 Garden chairs can be made from wood or plastic. Plastic is made from crude oil. Wood is obtained from trees. A student says that wooden chairs are better for sustainable development than plastic chairs.

10.1 Define the term sustainable development. **[1 mark]**

10.2 Write down **two** arguments that support what the student says. **[2 marks]**

10.3 Suggest **one** way in which plastic chairs are better than wooden chairs for sustainable development. **[1 mark]**

11 **Table 4** shows the boiling points of a series of hydrocarbons.

11.1 Identify the homologous series that the hydrocarbons in **Table 4** belong to. **[1 mark]**

11.2 Give the formula of the hydrocarbon in this homologous series with 9 carbon atoms. **[1 mark]**

11.3 Name the hydrocarbon with the chemical formula C_4H_{10}. **[1 mark]**

Table 4

Formula of hydrocarbon	Boiling point in °C
CH_4	−162
C_2H_6	−89
C_3H_8	−42
C_4H_{10}	−1
C_5H_{12}	36
C_6H_{14}	69
C_7H_{16}	98
C_8H_{18}	126

11.4 A student mixed some C_4H_{10} with bromine water. Give the student's observation. **[1 mark]**

11.5 The student carries out complete combustion with C_4H_{10}. Write the balanced symbol equation with state symbols for the reaction. **[3 marks]**

11.6 Another student carries out catalytic cracking on C_8H_{18}. Give the **two** conditions needed for catalytic cracking. **[2 marks]**

11.7 Draw a dot and cross diagram for the hydrocarbon CH_4. **[2 marks]**

11.8 CH_4 is a greenhouse gas. Explain how CH_4 in the atmosphere maintains the temperature on the Earth. **[4 marks]**

> **! Exam Tip**
>
> This is a covalent compound so needs overlapping circles.

12 A scientist carried out a series of tests to identify an unknown metal compound.

12.1 First the scientist wanted to identify the non-metal ion. They carried out three chemical tests. **Table 5** shows their observations.

Table 5

Chemical test	Observation
add barium chloride solution and hydrochloric acid	white precipitate formed
add silver nitrate solution and nitric acid	no precipitate
add dilute acid then collect and test gas	solution remained clear and colourless

Describe how the scientist tested the gas in the third chemical test. **[1 mark]**

12.2 Identify the non-metal ion in the metal compound. **[1 mark]**

12.3 To identify the metal, the scientist carried out a flame test on the metal compound. The scientist was unable to identify one colour in the flame test. Suggest why. **[1 mark]**

12.4 The scientist used flame emission spectroscopy to identify the metal in the compound. Give **three** advantages of using flame emission spectroscopy instead of a flame test to identify the metal. **[3 marks]**

12.5 **Figure 3** shows the results of the scientist's flame emission spectra and the spectra for five metals.

Figure 3

> **! Exam Tip**
>
> Use a ruler to compare the lines from the unknown metal with the given metals.

The scientist identified that the sample contains a mixture of three metal compounds. He identified one metal as sodium.

Identify the other **two** metals. **[2 marks]**

13 Nylon 6,6 is a polymer. The monomers that form the nylon 6,6 polymer are $H_2NC_6H_{12}NH_2$ and $ClCOC_4H_8COCl$.

13.1 **Figure 4** shows part of the structure of the nylon 6,6 polymer. Use the chemical formulae of the monomers to complete the polymer structure. **[2 marks]**

Figure 4

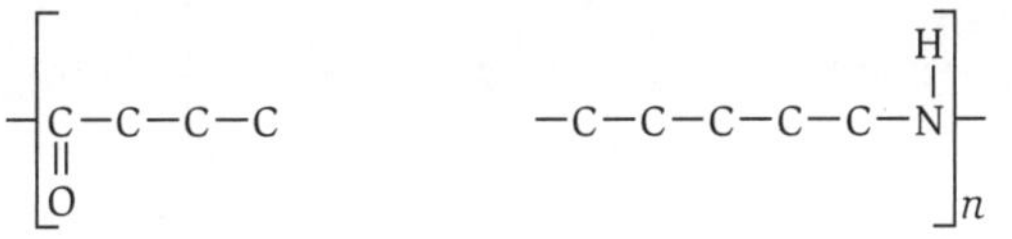

$$\left[\begin{array}{c} \text{C} - \text{C} - \text{C} - \text{C} \\ \| \\ \text{O} \end{array} \qquad \begin{array}{c} \text{H} \\ | \\ -\text{C} - \text{C} - \text{C} - \text{C} - \text{C} - \text{N} - \\ \end{array} \right]_n$$

13.2 A small molecule is produced when nylon 6,6 is formed. Name this type of polymerisation. **[1 mark]**

13.3 Identify the small molecule produced. **[1 mark]**

14 **Table 6** shows the percentages of the four most abundant gases in dry air.

Table 6

Gas	Percentage by volume in dry air
nitrogen	78.08
oxygen	20.95
argon	0.93
carbon dioxide	0.03

14.1 Which of these gases is believed to have been the most abundant in the Earth's early atmosphere? **[1 mark]**

14.2 Name one **other** gas that is believed to have been in the Earth's early atmosphere. **[1 mark]**

14.3 Which gas in **Table 6** is a greenhouse gas? **[1 mark]**

14.4 A student states that all greenhouse gases are bad. Is the student correct?

Give a reason for your answer. **[2 marks]**

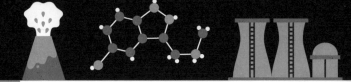

Knowledge

C20 Making our resources A

Corrosion

Corrosion is when a material reacts with substances in the environment and eventually wears away. Corrosion can be prevented in in two ways:

- physical barriers
- sacrificial protection

Rusting is an example of corrosion. It is caused by iron reacting with oxygen *and* water from the environment.

Physical barriers

The material is covered with a physical barrier like grease, paint, or a thin layer of another metal by a process called electroplating.

Aluminium reacts with oxygen to make a very thin layer of aluminium oxide around the metal that acts as a physical barrier. This layer then protects the rest of the metal from corrosion.

Sacrificial protection

A more reactive substance is placed on the material. The more reactive substance will react with the environment, and not the main material.

For example, iron is **galvanised** with zinc. The zinc then reacts with the oxygen and water in place of the iron.

Alloys

Alloys allow us to tailor the properties of metals to specific uses.

Alloy	Composition	Properties	Use
bronze	copper and tin	resistant to corrosion	statues, decorative items, ship propellers
brass	copper and zinc	very hard but workable	door fittings, taps, musical instruments
gold alloys	mostly gold with copper, silver and zinc added	attractive, corrosion resistant, hardness depends on carat	jewellery the proportion of gold is measured in carats. 24 carat gold contains 100% gold, 18 carat gold contains 75% gold
high carbon steel	iron with 1–2% carbon	strong but brittle	cutting tools, metal presses
low carbon steel	iron with <1% carbon	soft, easy to shape	extensive use in manufacture of cars, machinery, ships, containers, structural steel
stainless steel	iron with chromium and nickel	resistant to corrosion, hard	cutlery, plumbing
aluminium alloys	over 300 alloys available	low density, properties depend on composition	aircraft, military uses

Ceramics

Ceramics are materials with versatile properties that can have many different uses.

Ceramic	Manufacture	Properties	Uses
soda-lime glass	heat a mixture of sand, sodium carbonate, limestone	transparent, brittle	everyday glass objects
borosilicate glass	heat sand and boron trioxide	higher melting point than soda-lime glass	oven glassware, laboratory glassware
clay ceramics (pottery + bricks)	shape wet clay then heat in a furnace	hard, brittle, easy to shape before manufacture, resistant to corrosion	crockery, construction, plumbing fixtures

 **Key terms**

Make sure you can write a definition for these key terms.

alloy	ceramic	composite	corrosion	galvanise	matrix
reinforce	rusting	thermosetting	thermosoftening		

The properties of polymers depend on
- the monomers that make them up
- the conditions under which they are made.

For example, **low density poly(ethene)** and **high density poly(ethene)** are both made from ethene monomers but have very different properties due to the way that the polymer chains line up in the material.

Low density poly(ethene)

LDPE is formed when the addition polymerisation reaction of ethene is carried out under high pressure and in the presence of a small amount of oxygen.

The branched polymer chains cannot pack together, so causing the low density of the polymer.

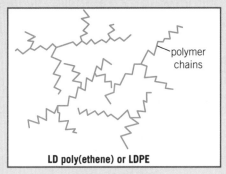

polymer chains

LD poly(ethene) or LDPE

High density poly(ethene)

HDPE is formed when the addition polymerisation reaction of ethene is carried out using a catalyst at 50 °C. The polymer chains are straight and can pack tightly together, so causing the high density of the polymer.

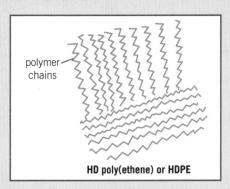

polymer chains

HD poly(ethene) or HDPE

Thermosoftening polymers

Thermosoftening polymers do not have links between the different chains, and soften when they are heated.

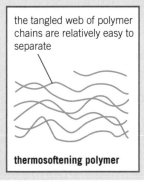

the tangled web of polymer chains are relatively easy to separate

thermosoftening polymer

Thermosetting polymers

Thermosetting polymers have strong links between the different chains, and do not melt when they are heated.

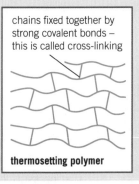

chains fixed together by strong covalent bonds – this is called cross-linking

thermosetting polymer

Composites

Composites are made from a main material (called a **matrix**) with fragments or fibres of other materials (called **reinforcements**) added into them. This means the material's properties can be made more useful.

Plywood and reinforced concrete are examples of composites.

Polymers

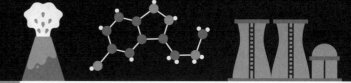

C20 Making our resources B

The Haber process

Fertilisers are important chemicals used to improve the growth of crop plants. Ammonia is a vital component of many fertilisers. It is produced in the **Haber process**:

- nitrogen + hydrogen ⇌ ammonia
- $N_2(g) + 2H_2(g) \rightleftharpoons 2NH_3(g)$

It is a reversible reaction, so the conditions affect the yield.

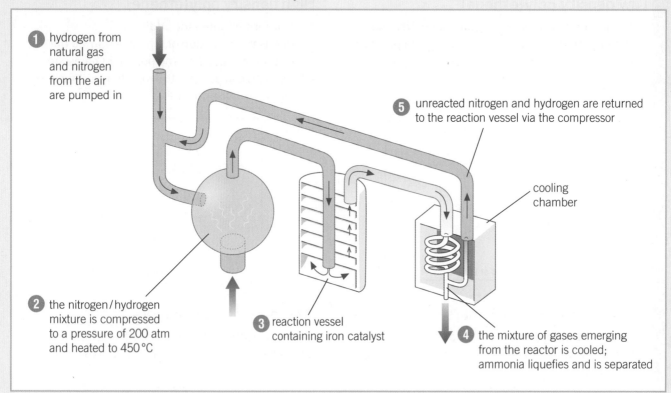

1 hydrogen from natural gas and nitrogen from the air are pumped in

5 unreacted nitrogen and hydrogen are returned to the reaction vessel via the compressor

cooling chamber

2 the nitrogen / hydrogen mixture is compressed to a pressure of 200 atm and heated to 450 °C

3 reaction vessel containing iron catalyst

4 the mixture of gases emerging from the reactor is cooled; ammonia liquefies and is separated

Conditions

Compromise

The conditions used for the Haber process are a *compromise* to balance yield, cost, and rate.

- an iron catalyst
- temperatures of about 450 °C
- pressure of about 200 atmospheres

Temperature

The forward reaction is exothermic. Therefore, lowering the temperature would increase the yield of ammonia, but would also decrease the rate of reaction.

Pressure

There are fewer gas molecules on the product side, so increasing the pressure would increase the yield and the rate of reaction. However, it is very expensive to increase the pressure.

Catalyst

Iron is an effective catalyst for the Haber process. It does not increase the yield, but does increase the rate.

Fertilisers

Fertilisers are produced industrially to increase the amount of food obtained from crops. Compounds containing nitrogen, phosphorous, and potassium are used, and fertilisers with all three in them are called **NPK fertilisers**.

NPK fertilisers are formulations. Some of the substances that go into NPK fertilisers can be mined straight from the ground (like potassium chloride or potassium sulfate). Others, like phosphate rock, need to be processed first. Phosphate rock can react with different acids to make different products, which can either be used as fertilisers on their own or added to an NPK fertiliser.

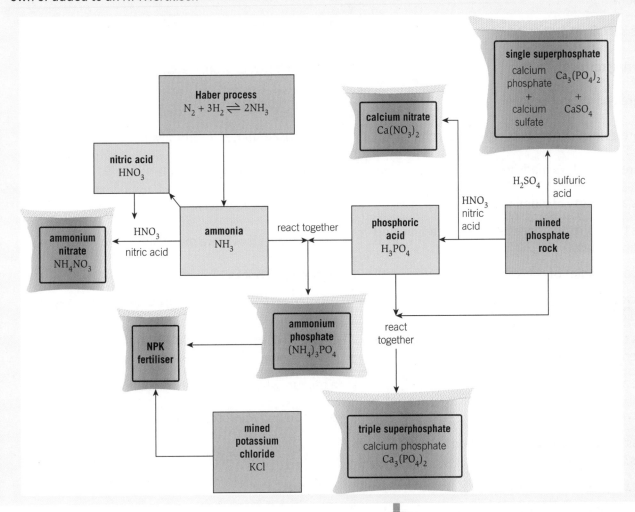

Laboratory vs. industry

The compounds found in fertilisers can be produced in the laboratory as well as industrially:

	laboratory	industrial
Quantities produced	small	large
Process	batch (do it once)	continuous (can keep doing it)
Apparatus	glass	stainless steel
Speed	slow	fast

 **Key terms**

Make sure you can write a definition for these key terms.

Haber process

NPK fertiliser

 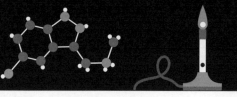 

Learn the answers to the questions below then cover the answers column with
a piece of paper and write down as many as you can. Check and repeat.

	C20 questions		Answers
1	What is corrosion?	Put paper here	the destruction of a material through reactions with substances in the environment
2	What physical barriers be used to protect against corrosion?		grease, paint, a thin layer of metal
3	What is sacrificial protection?		adding a more reactive metal to the surface of a material
4	How is rust formed?	Put paper here	reaction between iron, water, and oxygen
5	What are two alloys of copper?		brass and bronze
6	What are gold alloys in jewellery made from?		gold with copper, zinc, and silver
7	What are steel alloys made from?	Put paper here	iron, carbon, and other metals
8	What is a property of aluminium alloys?		generally have low densities
9	What is the main difference between soda-lime and borosilicate glass?		borosilicate glass has a much higher melting point
10	Give two examples of clay ceramics.	Put paper here	pottery and bricks
11	What two things do the properties of polymers depend on?		monomers and the conditions under which they are formed
12	What is the main difference between thermosetting and thermosoftening polymers?		thermosetting polymers do not soften when heated, thermosoftening polymers do
13	What is a composite?	Put paper here	a mixture of a matrix and reinforcements
14	Name two composites.		plywood and reinforced concrete
15	What is the balanced symbol equation for the Haber process?	Put paper here	$N_2(g) + 2H_2(g) \rightleftharpoons 2NH_3(g)$
16	What is the ammonia used for?		fertilisers
17	What is the effect of increasing the temperature of the Haber process on the yield, rate, and cost?		decrease yield, increase rate, increase cost
18	What is the effect of increasing the pressure of the Haber process on the yield, rate and cost?	Put paper here	increase yield, increase rate, increase cost
19	What catalyst do we use for the Haber process?		iron
20	What are the conditions for the Haber process?		450 °C, 200 atm, iron catalyst
21	What is an NPK fertiliser?		a formulation containing soluble compounds of nitrogen, phosphorous, and potassium

Now go back and use the questions below to check your knowledge from previous chapters

Previous questions / Answers

	Previous questions	Answers
1	In chemistry, what is a pure substance?	something made of only one type of substance
2	What are the four stages of a life cycle assessment?	• extracting and processing raw materials • manufacturing and packaging • use and operation during its lifetime • disposal at the end of its useful life
3	What is the atmosphere?	a layer of gas surrounding the Earth
4	What is the effect of decreasing the concentration of products on a reaction at dynamic equilibrium?	favours the forward reaction
5	What are the three reaction conditions that can be changed?	concentration, temperature, pressure
6	Why can scientists not be sure about the composition of the Earth's early atmosphere?	it was billions of years ago and evidence is limited
7	What is sustainable development?	development that meets the needs of current generations without compromising the ability of future generations to meet their own needs

Put paper here

Put paper here

Put paper here

Maths Skills

Practise your maths skills using the worked example and practice questions below.

Tangents

We can obtain the gradient of a curve at a specific point on a graph by drawing a tangent. A tangent is a straight line which touches the curve at only this specified point.

To do this, you draw a tangent line on the graph, then calculate the gradient of the tangent using:

$$\text{gradient} = \frac{\text{change in } y}{\text{change in } x}$$

Tangents can also be used to calculate the rate of a reaction at a given time on a curve.

Worked Example

Calculate the gradient of the line on the graph at 50 s.

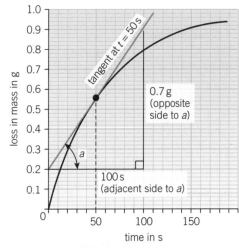

$$\text{Rate at 50 s} = \frac{0.7\,\text{g}}{100\,\text{s}} = 0.007\,\text{g/s}$$

(The gradient is the tangent of angle *a* in the right-angled triangle, i.e. opposite side divided by adjacent side.)

In this example, the gradient of the tangent is the same as the rate of the reaction at 50 s = 0.007 g/s

Practice

The graph below shows how the yield of ammonia in the Haber process changes depending on the temperature and the pressure.

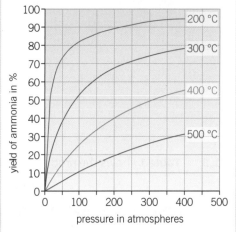

Which temperature has the steepest gradient at 50 atmospheres of pressure?

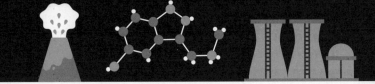

Practice

Exam-style questions

01 Most metals are used as alloys.

01.1 Explain why pure metals are alloyed. **[4 marks]**

_____

_____

_____

_____

01.2 Complete the sentences using answers from the box. Each word can be used once, more than once, or not at all. **[4 marks]**

| carbon | copper | gold | magnesium | tin | zinc |

Bronze is an alloy that is made from _____ and

_____ . Brass is an alloy made from _____ and

_____ .

01.3 Gold is often used to make jewellery.

Suggest why jewellery made from gold is more valuable then jewellery made from a similar coloured metals like copper. **[1 mark]**

_____

_____

01.4 The purity of a sample of gold is measured in carats. 24 carat gold is pure gold. 18 carat gold is an alloy made of 75% gold mixed with other materials.

Calculate the percentage of gold in a 14 carat gold sample. **[2 marks]**

> **! Exam Tip**
>
> You probably haven't come across a question like this in this exact context but you need to get used to applying the maths you know in new situations.

_____ %

01.5 **Table 1** shows the properties of three alloys.

Table 1

Alloy	Properties	Composition
alloy **A**	strong, brittle	0.6% carbon, 0.5% manganese, iron
alloy **B**	hard, resistant to corrosion	20% chromium, 15% nickel, iron
alloy **C**	soft, easily shaped	0.05% carbon, 0.1% manganese, iron

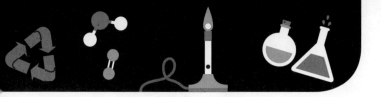

Use the information in **Table 1** to identify the three alloys. **[3 marks]**

alloy **A** _____

alloy **B** _____

alloy **C** _____

02 PVA glue contains the polymer poly(vinyl alcohol). **Figure 1** shows part of the structure of poly(vinyl alcohol).

Figure 1

$$
\begin{array}{cccccccc}
H & H & H & H & H & H & H & H \\
| & | & | & | & | & | & | & | \\
-C & -C & -C & -C & -C & -C & -C & -C- \\
| & | & | & | & | & | & | & | \\
H & OH & H & OH & H & OH & H & OH \\
\end{array}
$$

02.1 Draw the monomer of poly(vinyl alcohol). **[1 mark]**

02.2 Identify the type of polymerisation that forms poly(vinyl alcohol). **[1 mark]**

_____

_____

02.3 PVA glue is a thick liquid. When the chemical borax is added to PVA glue, the polymer shown in **Figure 2** is formed.

Figure 2

$$
\begin{array}{cccccccc}
H & H & H & H & H & H & H & H \\
| & | & | & | & | & | & | & | \\
-C & -C & -C & -C & -C & -C & -C & -C- \\
| & | & | & | & | & | & | & | \\
H & O & H & O & H & O & H & O \\
& \diagdown & & \diagdown & & \diagdown & & \diagdown \\
& & B & & & & B & \\
& \diagup & & \diagup & & \diagup & & \diagup \\
H & O & H & O & H & O & H & O \\
| & | & | & | & | & | & | & | \\
-C & -C & -C & -C & -C & -C & -C & -C- \\
| & | & | & | & | & | & | & | \\
H & H & H & H & H & H & H & H \\
\end{array}
$$

Predict how the properties of the polymer in **Figure 2** would differ from poly(vinyl alcohol). Explain your answer. **[3 marks]**

_____

_____

_____

_____

_____

03 The Haber process is used to produce ammonia.

03.1 Give the balanced symbol equation for the reaction. **[2 marks]**

03.2 Give **one** use for the ammonia produced in the Haber process. **[1 mark]**

03.3 Give **one** source of the nitrogen in the Haber process. **[1 mark]**

03.4 Give the conditions used in the industrial Haber process. Explain how the conditions make the Haber process economically viable in an industrial setting. **[6 marks]**

04 This question is about NPK fertilisers.

04.1 Name the **three** elements found in NPK fertilisers. **[1 mark]**

04.2 Give **one** advantage and **one** disadvantage for a farmer putting NPK fertiliser on their fields. **[4 marks]**

04.3 Ammonia is produced in a laboratory by heating ammonium chloride and calcium hydroxide using the experimental setup shown in **Figure 3**.

Figure 3

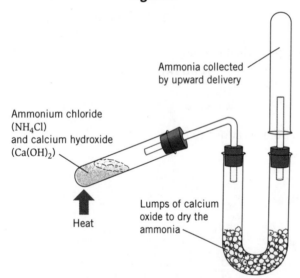

Ammonia collected by upward delivery

Ammonium chloride (NH_4Cl) and calcium hydroxide ($Ca(OH)_2$)

Heat

Lumps of calcium oxide to dry the ammonia

Balance the symbol equation for the reaction.

_____NH_4Cl + $Ca(OH)_2$ → _____NH_3 + $CaCl_2$ + _____H_2O

04.4 Describe how the industrial production of ammonia is different to the laboratory production of ammonia. **[4 marks]**

05 Steel is an alloy of iron. Compared to iron, stainless steel is resistant to corrosion.

05.1 Define the term corrosion. **[1 mark]**

05.2 Give **one** other property of stainless steel compared to pure iron. **[1 mark]**

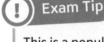

Exam Tip

When writing equations, start by writing down the chemical formula of the three substances, then balance the equation.

Exam Tip

This is a popular question as it ties in lots of chemistry from other units.

Exam Tip

You can use the Periodic Table to help you find the three elements if you can't remember them.

05.3   To prevent pure iron from corroding, a barrier can be used.
Explain how a barrier protects iron from corrosion.   **[2 marks]**

05.4   Explain why aluminium does not need to have a barrier added.   **[3 marks]**

06   This question is about the corrosion of metals.

06.1   Rust is also known as iron(III) hydroxide.
Complete the symbol equation for the rusting of iron.   **[2 marks]**

$4Fe + \rule{1.5cm}{0.4pt} + \rule{1.5cm}{0.4pt} H_2O \rightarrow \rule{1.5cm}{0.4pt} Fe(OH)_3$

06.2   A student wanted to run an experiment to see which factors affect corrosion. They decided to test the following five conditions.

A: no water, no air   **B:** water, no air   **C:** air, no water

D: water and air   **E:** salt water and air

The method used is as follows:

1   Label five test tubes **A**, **B**, **C**, **D**, and **E**.

2   Put a nail into each test tube.

3   Half-fill test tubes **B**, **D**, and **E** with water.

4   Put a spoon full of salt into test tube **E**.

5   Put a lid on test tubes **A** and **B**.

6   Leave to rust.

Describe **three** ways the method could be improved to give more accurate results.   **[6 marks]**

> **! Exam Tip**
>
> Describe means state what is wrong and then how you could improve it. Your answer should have three problems followed by fixes.

06.3   Predict which conditions will cause the nail to rust the most. Explain your answer.   **[3 marks]**

06.4   Zinc is used as a sacrificial metal to stop iron from corroding. Describe how zinc prevents iron from rusting.   **[2 marks]**

07   Glass is a composite material made mainly from sand and other compounds. Two types of glass are soda-lime glass and borosilicate glass.

07.1   Name the substances that borosilicate glass is made from.   **[1 mark]**

07.2   Give **one** difference in the properties of soda-lime glass and borosilicate glass.   **[1 mark]**

07.3   Soda-lime glass is made from sand, sodium carbonate, and limestone. Give the formula for sodium carbonate.   **[1 mark]**

> **! Exam Tip**
>
> It is important that you know the formulae of all the common ions and how to combine them to make ionic compounds. To work out the charge of a metal ion in Group 1 or Group 2, look at the number of electrons in its outer shell.

07.4 The properties of glass will depend on how it is produced.

- Crystal glass is made from sand and lead oxide. It is very fragile but is very clear and shines brightly.

- Aluminosilicate glass is made from sand and aluminium oxide. It is very resistant to weathering and corrosion.

- Germanium oxide glass is made from sand and germanium dioxide. It is very clear and does not reduce the intensity of light passed through it.

For each of the three functions, choose the most appropriate glass. **[3 marks]**

- fibre optic cables
- decorative vase
- windows

Exam Tip

There is a lot of information in this question. Use highlighters to colour-code the information.

08 Both low density poly(ethene) (LDPE) and high density poly(ethene) (HDPE) are made from the same starting materials.

08.1 Draw the repeating unit of poly(ethene). **[2 marks]**

08.2 Draw the monomer that forms poly(ethene). **[1 mark]**

08.3 Compare the structure of LDPE and HDPE **[4 marks]**

08.4 Both LDPE and HDPE are thermosoftening polymers.
Explain why LDPE and HDPE are thermosoftening polymers.

Exam Tip

Remember, the monomer will be the substance thats within the brackets of the polymer name. In this instance, it's ethene.

09 This question is about NPK fertilisers.

09.1 Name the salt produced when phosphate rock reacts with nitric acid. **[1 mark]**

09.2 Phosphoric acid, H_3PO_4, reacts with phosphate rock to form calcium phosphate. Calcium has a 2+ charge.
Give the chemical formula of calcium phosphate. **[1 mark]**

09.3 Phosphate rock reacts with an acid to form calcium phosphate and calcium sulfate. Identify the acid. **[1 mark]**

09.4 Ammonium nitrate, NH_4NO_3, is another salt that is used in NPK fertilisers.
Describe fully how ammonium nitrate is produced. **[4 marks]**

10 Phosphate can be mined form rock. Phosphate rock is the general name given to any rock that has high levels of phosphate. One such rock was found to have the chemical formula $X_5Cl(PO_4)_3$, where X is a metal. The relative atomic mass of $X_5Cl(PO_4)_3$ is 520.5.

10.1 Determine the identity of X. **[3 marks]**

10.2 Identify the charge on X. **[1 mark]**

Exam Tip

Be careful with the number of phosphorous and oxygen atoms.

10.3 The phosphate rock is treated with nitric acid to produce phosphoric acid.

Name the other product of this reaction. **[1 mark]**

10.4 Phosphoric acid can be neutralised with ammonia to give ammonium phosphate.

Give the formula of ammonium phosphate. **[1 mark]**

> **(!) Exam Tip**
>
> You need to be able to recall the names of the salts produced when phosphate rock is treated with different acids.

11 Ammonia is an important chemical that is produced industrially in the Haber process. **Figure 4** shows how the percentage yield of ammonia changes with different reaction conditions.

Figure 4

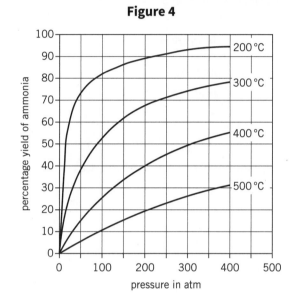

11.1 Write the balanced symbol equation with state symbols for the reaction in the Haber process. **[3 marks]**

11.2 Sketch a data line on **Figure 3** for the temperature that the Haber process is carried out at industrially. **[1 mark]**

11.3 Use your data line from **11.2** to identify the expected percentage yield of ammonia in the industrial Haber process. **[1 mark]**

11.4 Use **Figure 3** and your knowledge of the Haber process to explain why the industrial conditions of the Haber process are described as a compromise. **[6 marks]**

11.5 Describe the role of ammonia in food production. **[2 marks]**

12 Most metals are found bonded to other elements in compounds within rocks. The metal needs to be extracted from the compounds.

12.1 Explain why scientists are developing new techniques to extract metals from their ores. **[3 marks]**

> **(!) Exam Tip**
>
> You've been asked to give the state symbols, so don't forget them! Many students do, and this is an easy mark.

12.2 Evaluate the use of phytomining to obtain copper. **[6 marks]**

12.3 **Table 2** shows three plants that are used in phytomining.

Table 2

Plant	Mass of copper produced per 1 kg of biomass in mg	Preferred climate	Growth
A	17000	hot humid forests	fast
B	5000	cold dry desert	slow
C	2000	moderate and damp	fast

Use **Table 2** to identify the best crop to use to extract copper from a field in the UK.

Justify your answer. **[4 marks]**

12.4 Calculate the mass in kg of copper mined from 14.5 kg of plant **B**.
1000 mg = 1 g **[2 marks]**

> **! Exam Tip**
>
> Think about your knowledge of the process of phytomining, and then bring in bring in topics from other parts of your chemistry course to answer this question.

13 A student wanted to look at the different compounds in a mixture. The student set up the paper chromatography experiment in **Figure 5** to separate out the compounds.

Figure 5

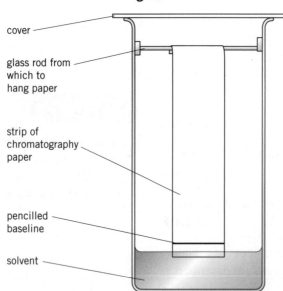

cover

glass rod from which to hang paper

strip of chromatography paper

pencilled baseline

solvent

13.1 Why did the student draw the start line in pencil and not in pen? **[1 mark]**

13.2 Why did the student ensure the solvent was below the start line? **[1 mark]**

13.3 Why did the student place a lid on top of the beaker? **[1 mark]**

13.4 The mixture was made of three compounds. Compound **A** is a pure substance. Compound **B** is a mixture of **A** and **C**. Compound **C** is a pure substance that is different to **A**. Sketch the chromatogram that the student produced. **[3 marks]**

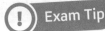 **Exam Tip**

Sketch means you don't need to include values but you do still need to use a ruler and label everything.

13.5 The R_f value can be used to determine the identity of a substance in chromatography.

Give the equation to calculate the R_f value of a substance. **[1 mark]**

13.6 The student calculated the R_f value of substance **C** to be 0.31. Use **Table 3** to identify substance **C**. **[1 mark]**

 Exam Tip

None of the values in the table exactly match the value calculated by the student, but often an experimentally-obtained value won't be identical to the value from a database.

Table 3

Substance	R_f value
methyl red	0.30
ethyl green	0.46
titan yellow	0.61
fuchsin acid	0.89

14 **Figure 6** shows four chemical structures.

Figure 6

A

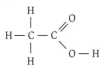

B

C=C structure with H, H on left carbon and H, H on right carbon

C

HOOC—CH₂—CH₂—CH₂—CH₂—COOH

D

HO—CH₂—CH₂—OH

14.1 Which chemical structure is a diol? **[1 mark]**

14.2 Which chemical structure is an alkene? **[1 mark]**

14.3 Identify which **two** structures will form a polyester. **[1 mark]**

14.4 Identify which structure will form an addition polymer. Name the polymer formed. **[2 marks]**

14.5 Draw the polymer formed from the polymerisation of HO—CH₂—OH and HOOC—CH₂—COOH. **[2 marks]**

 **Exam Tip**

Identify the functional groups, then think about the small molecule lost and where it might come from.

key

relative atomic mass
atomic symbol
name
atomic (proton) number

Example:

1
H
hydrogen
1

Group 1	2											3	4	5	6	7	0
																	4 **He** helium 2
7 **Li** lithium 3	9 **Be** beryllium 4											11 **B** boron 5	12 **C** carbon 6	14 **N** nitrogen 7	16 **O** oxygen 8	19 **F** fluorine 9	20 **Ne** neon 10
23 **Na** sodium 11	24 **Mg** magnesium 12											27 **Al** aluminium 13	28 **Si** silicon 14	31 **P** phosphorus 15	32 **S** sulfur 16	35.5 **Cl** chlorine 17	40 **Ar** argon 18
39 **K** potassium 19	40 **Ca** calcium 20	45 **Sc** scandium 21	48 **Ti** titanium 22	51 **V** vanadium 23	52 **Cr** chromium 24	55 **Mn** manganese 25	56 **Fe** iron 26	59 **Co** cobalt 27	59 **Ni** nickel 28	63.5 **Cu** copper 29	65 **Zn** zinc 30	70 **Ga** gallium 31	73 **Ge** germanium 32	75 **As** arsenic 33	79 **Se** selenium 34	80 **Br** bromine 35	84 **Kr** krypton 36
85 **Rb** rubidium 37	88 **Sr** strontium 38	89 **Y** yttrium 39	91 **Zr** zirconium 40	93 **Nb** niobium 41	96 **Mo** molybdenum 42	[98] **Tc** technetium 43	101 **Ru** ruthenium 44	103 **Rh** rhodium 45	106 **Pd** palladium 46	108 **Ag** silver 47	112 **Cd** cadmium 48	115 **In** indium 49	119 **Sn** tin 50	122 **Sb** antimony 51	128 **Te** tellurium 52	127 **I** iodine 53	131 **Xe** xenon 54
133 **Cs** caesium 55	137 **Ba** barium 56	139 **La*** lanthanum 57	178 **Hf** hafnium 72	181 **Ta** tantalum 73	184 **W** tungsten 74	186 **Re** rhenium 75	190 **Os** osmium 76	192 **Ir** iridium 77	195 **Pt** platinum 78	197 **Au** gold 79	201 **Hg** mercury 80	204 **Tl** thallium 81	207 **Pb** lead 82	209 **Bi** bismuth 83	[209] **Po** polonium 84	[210] **At** astatine 85	[222] **Rn** radon 86
[223] **Fr** francium 87	[225] **Ra** radium 88	[227] **Ac*** actinium 89	[261] **Rf** rutherfordium 104	[262] **Db** dubnium 105	[266] **Sg** seaborgium 106	[264] **Bh** bohrium 107	[277] **Hs** hassium 108	[268] **Mt** meitnerium 109	[271] **Ds** darmstadtium 110	[272] **Rg** roentgenium 111	[285] **Cn** copernicium 112	[286] **Nh** nihonium 113	[289] **Fl** flerovium 114	[289] **Mc** moscovium 115	[293] **Lv** livermorium 116	[294] **Ts** tennessine 117	[294] **Og** oganesson 118

*The lanthanides (atomic numbers 58–71) and the actinides (atomic numbers 90–103) have been omitted.

Relative atomic masses for **Cu** and **Cl** have not been rounded to the nearest whole number.

Great Clarendon Street, Oxford, OX2 6DP, United Kingdom

Oxford University Press is a department of the University of Oxford.

It furthers the University's objective of excellence in research, scholarship, and education by publishing worldwide. Oxford is a registered trade mark of Oxford University Press in the UK and in

certain other countries

British Library Cataloguing in Publication Data

Data available

978-1-38-200485-5

10 9 8 7 6 5 4 3 2 1

Paper used in the production of this book is a natural, recyclable product made from wood grown in sustainable forests.

The manufacturing process conforms to the environmental regulations of the country of origin.

Printed in China

Acknowledgements

The publisher would like to thank the following for permissions to use copyright material:

Philippa Gardom Hulme would like to thank Mary and Edward Hulme, and Sarah, Catherine and Barney Gardom.

Cover illustration: Andrew Groves

p14: Shutterstock; **p88**: SCIENCE PHOTO LIBRARY.

Other artwork by Q2A Media Services Inc. and OUP.

Although we have made every effort to trace and contact all copyright holders before publication this has not been possible in all cases. If notified, the publisher will rectify any errors or omissions at the earliest opportunity.